Limits, Limits Everywhere

Photo of the author standing in front of Richard Dedekind's house in Braunschweig (July 2010) © Tijana Levajkovic

Limits, Limits Everywhere

The Tools of
Mathematical Analysis

DAVID APPLEBAUM

OXFORD
UNIVERSITY PRESS

OXFORD
UNIVERSITY PRESS

Great Clarendon Street, Oxford OX2 6DP

Oxford University Press is a department of the University of Oxford.
It furthers the University's objective of excellence in research, scholarship,
and education by publishing worldwide in

Oxford New York

Auckland Cape Town Dar es Salaam Hong Kong Karachi
Kuala Lumpur Madrid Melbourne Mexico City Nairobi
New Delhi Shanghai Taipei Toronto

With offices in

Argentina Austria Brazil Chile Czech Republic France Greece
Guatemala Hungary Italy Japan Poland Portugal Singapore
South Korea Switzerland Thailand Turkey Ukraine Vietnam

Oxford is a registered trade mark of Oxford University Press
in the UK and in certain other countries

Published in the United States
by Oxford University Press Inc., New York

British Library Cataloguing in Publication Data
Data available

Library of Congress Cataloging in Publication Data
Library of Congress Control Number: 2011945216

Typeset by Cenveo Publisher Services
Printed in Great Britain
on acid-free paper by
Clays Ltd, St Ives plc

ISBN 978-0-19-964008-9

1 3 5 7 9 10 8 6 4 2

To Ben and Kate 'the cool.'

Strong stuff, isn't it? He paused. Limits, limits everywhere.
William Boyd, *The New Confessions*

Introduction

This book is written for anyone who has an interest in mathematics. It occupies territory that lies midway between a popular science book and a traditional textbook. Its subject matter is the part of mathematics that is called *analysis*. This is a very rich branch of mathematics that is also relatively young in historical terms. It was first developed in the nineteenth century but it is still very much alive as an area of contemporary research. Analysis is typically first introduced in undergraduate mathematics degree courses as providing a 'rigorous' (i.e. logically flawless) foundation to a historically older branch of mathematics called the *calculus*. Calculus is the mathematics of motion and change. It evolved from the work of Isaac Newton and Gottfried Leibniz in the seventeenth century to become one of the most important tools in applied mathematics. Now the relationship between analysis and calculus is extremely important but it is not the subject matter of this book. Indeed readers do not need to know any calculus at all to read most of it.

So what is this book about? In a sense it is about two concepts – *number* and *limit*. Analysis provides the tools for understanding what numbers really are. It helps us make sense of the infinitesimally small and the infinitely large as well as the boundless realms between them. It achieves this by means of a key concept – the limit – which is one of the most subtle and exquisitely beautiful ideas ever conceived by humanity. This book is designed to gently guide the reader through the lore of this concept so that it becomes a friend.

So who is this book for? I envisage readers as falling into one of three (not necessarily disjoint) categories:

- The curious. You may have read a popular book on mathematics by a masterful expositor such as Marcus de Sautoy or Ian Stewart. These books stimulated you and started you thinking. You'd like to go further but don't have the time or background to take a formal course – and self-study from a standard textbook is a little forbidding.
- The confused. You are at university and taking a beginner's course in analysis. You are finding it hard and are seriously lost. Maybe this book can help you find your way?
- The eager. You are still at school. Mathematics is one of your favourite subjects and you love reading about it and discovering new things. You've picked up this book in the hope of finding out more about what goes on at college/university level.

To read this book requires some mathematical background but not an awful lot. You should be able to add, subtract, multiply and divide whole numbers and fractions. You should also be able to work with school algebra at the symbolic level. So you need to be able to calculate fractions like $\frac{1}{2} - \frac{1}{3} = \frac{1}{6}$ and also be able to deal with the general case $\frac{a}{b} - \frac{c}{d} = \frac{ad - bc}{bd}$. I'll take it for granted that you can multiply brackets to get $(x + y)(a + b) = xa + ya + xb + yb$ and also recognise identities such as the 'difference of two squares' $x^2 - y^2 = (x + y)(x - y)$. Beyond this it is vital that you are willing to allow your mind to engage with extensive bouts of systematic logical thought.

As I pointed out above, you don't need to know anything about calculus to read most of this book (but anything you do know can only help). To keep things as simple as possible, I avoid the use of set theory (at least until the end of the book) and the technique of 'proof by mathematical induction,' but both topics are at least briefly introduced in appendices for those who would like to become acquainted with them.

Sometimes I am asked what mathematicians really do. Of course there are as many different answers as there are many different traditions within the vast scope of modern mathematics. But an essential feature of what is sometimes called 'pure' mathematics is the process of 'proving theorems'. A theorem is a fancy way of talking about a chunk of mathematical knowledge that can be expressed in two or three sentences and that tells you something new. A proof is the logical argument we use to convince ourselves (and colleagues, students and readers) that this new knowledge is really correct. If you pick up a mathematics book in a library it may well be that 70 to 80% of it just consists of lists of theorems and proofs – one after the other. On the other hand most expository books about mathematics that are written for a general reader will contain none of these at all. In this book you'll find a halfway house. The author's goal is to give the reader a genuine insight into how mathematicians really think and work. So you're going to meet theorems and proofs – but the development of these is going to be very gentle and easy paced. Each time there will be discussion before and after and – at least in the early part of the book, when the procedures are unfamiliar – the proofs will be spelt out in much greater detail than would be the case in a typical textbook.

So what is the book about anyway? In once sense it is the story of a quest – the long quest of the human race to understand the notion of number. In a sense there are two types of number. There are those like the whole numbers 1, 2, 3, 4, 5, 6 etc. that come in discrete chunks and there are those that we call 'real numbers' that form a continuum where each successive number merges into the last and there are no gaps between them.[1] It is this second type of

---

[1] This is an imprecise suggestive statement. If you want to perceive the truth that lies behind it then you must read the whole book.

number that is the true domain of analysis.[2] These numbers may appear to be very familiar to us and we may think that we 'understand' them. For example you all know the number that we signify by π. It originates in geometry as the universal constant you obtain when the circumference of any (idealised) circle is divided by its diameter. You may think you know this number because you can find it on a pocket calculator (mine gives it as 3.1415927) but I hope to be able to convince you that your calculator is lying to you. You really don't know π at all – and neither do I. This is because the calculator only tells us part of the decimal expansion of π (with enough accuracy to be fine for most practical applications) but the 'true' decimal expansion of π is *infinite*. We are only human beings with limited powers and our brains are not adapted to grasp the infinite as a whole. But mathematicians have developed a tool which enables us to gain profound insights into the infinite nature of numbers by only ever using *finite* means. This tool is called the limit and this book will help you understand how it works.

Guide for Readers

There are thirteen chapters in this book which is itself divided into two parts. Part I comprises Chapters 1 to 6 and Part II is the rest. The six chapters in Part I can serve as a background text for a standard first year undergraduate course in numbers, sequence and series (or in some colleges and universities, the first half of a first or second year course on introductory real analysis). Chapters 1 and 2 introduce the different types of number that feature in most of this book: natural, prime, integer, rational and real. Chapter 3 is the bridge between number and analysis. It is devoted to the art of manipulating inequalities. Chapters 4 and 5 focus on limits of sequences and begin the study of analysis proper. Chapter 6 (which is the longest in the book) deals with infinite series.

Part II comprises a selection of additional interesting topics. In Chapter 7 we meet three of the most fascinating numbers in mathematics: e, π and γ. Chapters 8 and 9 introduce two topics that normally don't feature in standard undergraduate courses – infinite products and continued fractions (respectively). Chapter 10 begins the study of the remarkable theory of infinite numbers. Chapter 11 is perhaps, from a conceptual perspective, the most difficult chapter in the book as it deals with the rigorous construction of the real numbers using Dedekind cuts and the vital concept of completeness. Chapter 12 is a rapid survey of the further development of analysis into the realm of functions, continuity and the calculus.

---

[2] To be precise *real analysis*, which is that part of analysis which deals with real sequences, series and functions. This topic should be distinguished from *complex analysis* which studies complex sequences, series and functions and which isn't the subject of this book, although we do briefly touch on it in Chapter 8.

In Chapter 13 we give a brief account of the history of the subject and in Further Reading we review some of the literature that the reader might turn to next after reading this book.

You learn mathematics by doing and not by reading and so each chapter in Part I closes with a set of exercises which you are strongly encouraged to attempt.[3] As well as practising techniques, these also enable you to further develop some aspects of the theory that are omitted from the main text (where explicit guidance is generally provided). So for example, in the exercises for Chapters 4 and 5 you meet the useful concept of a subsequence and can prove the Bolzano–Weierstrass theorem for yourself. Hints and solutions to selected exercises can be found at the end of the book. Professional educators can obtain full solutions by following instructions that can be found on http:// ukcatalogue.oup.com/product/9780199640089.do

I would expect that most readers will have ready access to the internet and so I have included a lot of references to Wikipedia throughout the text. This is so that you can very quickly find out a lot more about a topic if you want to. However bear in mind that Wikipedia is not yet thoroughly reliable and you should never quote mathematical results found there unless you have also confirmed them by consulting an authoritative text.

Note for Professional Educators

As pointed out above, the book falls naturally into two parts. Part I shadows a first year university mathematics course on sequences and series but with a little bit of number theory thrown into the mix. No calculus is used in Part I except at the very end in an optional aside. There is also no set theory until the very end of the book. Part II is a collection of subsidiary topics. I feel freer to use calculus here – but only occasionally. Indeed elementary analytic number theory takes over some of the usual role of calculus in this book as motivation for learning analysis. I avoid the axiomatic method throughout this book. This is a pedagogic device rather than a philosophical standpoint. I believe it is more helpful for those encountering the properties of real numbers for the first time to first develop basic analytical insight into their manipulation. The niceties of complete ordered fields can then be left to a later stage of their education. Indeed as Ivor Grattan-Guinness writes in 'The Rainbow of Mathematics' (p.740), 'The teaching of axioms should come *after* conveying the theory in a looser version'.

---

[3] These are drawn from a variety of sources including some of the textbooks listed in Further Reading.

Acknowledgements

I would like to thank Geoff Boden and Paul Mitchener who both read through draft versions and gave me very helpful feedback. In addition the anonymous referees gave highly valuable input. Finally a big thanks to all at Oxford University Press who helped turn this project into the book you are now reading – particularly my editors Beth Hannon and Keith Mansfield.

Contents

Part I

Approaching Limits

1

A Whole Lot of Numbers

"Think about maths Felix," Levin advised seriously as I walked out,
"it'll take your mind off your nervous breakdown."
White Light, Rudy Rucker

1.1 Natural Numbers

The numbers that we learn to count with are called *natural numbers* by
mathematicians. If we try to make a list of them that starts with the smallest
we begin 1, 2, 3, 4, 5, 6, 7, 8, 9, 10, 11, 12, 13, ... and then at some stage we get
bored and so we stop, writing the three dots ... to indicate 'etcetera' or 'so it goes'.
I stopped at 13 but there is no good reason to do this. We can go on to 100 or 127
or 1000 or 1000000 or any enormously large number we care to choose. There are
some impressively big numbers out there. At the time of writing the population of
the world is estimated to be 6762875008,[1] and some predictions expect it to reach
around 9 billion by 2040.[2]

This pales into insignificance when compared to some of the numbers we
can write down such as 10^{100} which is one followed by a hundred noughts. This
number is sometimes called a *googol*. An even larger number is $10^{10^{100}}$ which is
one followed by a googol of noughts and this is called a *googolplex*. It's easy to
create large numbers in this way but this is a process that has no end. There is no
largest natural number and this leads us to use phrases like 'the numbers go on to
infinity'. What do we mean by 'infinity' when we make such a statement? Are we
indicating some mysterious concept that lies beyond our usual understanding of

---

[1] See http://www.census.gov/ipc/www/idb/worldpopinfo.html
[2] See http://en.wikipedia.org/wiki/World_population

numbers? To probe further it would be useful to give a precise argument which makes it logically clear that there cannot be a largest natural number. This uses a technique that mathematicians call *proof by contradiction* which we will employ time and time again in this book, so it's a good idea to get used to it as soon as possible. We begin by making an assertion that we intend to disprove. In this case it is 'there is a largest number'. Let's give this largest number a symbol N. Now suppose that N is a legitimate natural number. Then we can add 1 to it to get another number $N + 1$. Now there are three possibilities for relating N to $N + 1$. Either $N + 1$ is larger, equal to or smaller than N. Now if $N + 1$ is larger than N then we've created our desired contradiction as we have a number that is larger than the largest and this cannot be. If $N + 1 = N$ then we can subtract N from both sides to get $1 = 0$ which is also a contradiction.[3] I'll leave you to work out for yourself the contradiction that follows from supposing that $N + 1$ is smaller than N. In all cases we see that if a number such as N exists then we have a contradiction. As contradictions are not allowed in mathematics, we conclude that a largest natural number cannot exist. When we say the 'numbers go on to infinity' we are really doing nothing more than describing the fact that counting can never end. From this point of view 'infinity' is not some mystical unreachable end-point but a linguistic label we use to indicate a never-ending process.

1.2 Prime Numbers

We can add natural numbers together to make new numbers to our heart's content. We can also subtract b from a to make the natural number $a - b$ but this only works if b is smaller than a. Addition and subtraction are *mutually inverse* to each other in that they undo the effect of each other. This vague wording is made more precise with symbols: $(a + b) - b = (a - b) + b = a$ where the operation in brackets is always carried out first. Repeated addition of the same number to itself is simplified by introducing multiplication so $a + a = 2 \times a$ which it is more convenient to write as $2a$, and more generally if we add a to itself n times then we get na. Just as subtraction and addition are mutually inverse then so are multiplication and division. Indeed we say that $b \div a = n$ or equivalently $\frac{b}{a} = n$ provided that $b = na$. It's important to be aware that at this stage we are only dealing with natural numbers and so $b \div a$ is only defined for our present purposes if there is no remainder when division is carried out — so $6 \div 3 = 2$ but $5 \div 3$ is not allowed. Now suppose that we have a number b than can be written $b = na$ so that $b \div a = n$ and $b \div n = a$. We say that a and n are *factors* (also

---

[3] 0 is not a natural number but this doesn't invalidate the argument.

1 A WHOLE LOT OF NUMBERS

called *divisors*) of *b*, e.g. 2 and 3 are factors of 6 and 10 and 7 are factors of 70. More generally we can see that 6 is 'built' from its factors by multiplying them together. Now we'll ask a very important question.

Can we find a special collection of natural numbers that has the property that every other number can be built from them by multiplying a finite number of these together (with repetitions if necessary)?

We will see that the answer to this question is affirmative and that the numbers that we need are the *prime numbers* or *primes*. A formal definition of a prime number is that it is a natural number that is greater than 1 whose only factors are 1 and itself. So if *p* is prime we have $p = 1 \times p = p \times 1$ but there are no other numbers *a* and *n* such that $p = na$. The list of prime numbers begins

2, 3, 5, 7, 11, 13, 17, 19, 23, 29, 31, 37, 41, 43, 47, 53, 59, 61, 67, 71, 73, 79, 83, 89, 97, 101, ...

Note that 2 is the only even prime number (why?) Prime numbers are on the one hand, rather simple things but on the other hand they are one of the most mysterious and intractable mathematical objects that we have ever discovered. First of all there is no clear discernible pattern in the succession of prime numbers – there is no magic formula which we can use to find the *n*th prime. At the start of the list of natural numbers, prime numbers seem quite frequent but as we progress to higher and higher numbers then they appear to be rarer and rarer. Later on in this section we will demonstrate two interesting facts. Firstly that there are an infinite number of prime numbers (and so there is no largest prime) and secondly given any natural number *m* we can (if we start at a large enough number) find $m + 1$ consecutive natural numbers, none of which is prime.

Before we explore these ideas, we'll return to the question that motivated us to look at prime numbers in the first place.

A natural number that is bigger than 1 and isn't prime is said to be *composite*. So if *b* is composite we can always find two factors *a* and *c* such that $b = ac$. Consider the number 720. It is clearly composite as $720 = 2 \times 360$ or 10×72. Let's look more closely at the second of these products. As $10 = 2 \times 5$ and $72 = 6 \times 12 = (2 \times 3) \times (3 \times 2 \times 2)$, after rearranging we can write $720 = 2 \times 2 \times 2 \times 2 \times 3 \times 3 \times 5 = 2^4 3^2 5$. Now 2, 3 and 5 are all prime and the decomposition we've found is called the *prime factorisation* of 720. We've built it by multiplying together (as many times as were needed) all the prime numbers that are factors of our number. We call these numbers *prime factors* as they are both prime and also are factors of the number in question. Now suppose that we are given a general natural number *n* whose prime factors are $p_1, p_2, \ldots, p_N$. We will write $n = p_1^{m_1} p_2^{m_2} \cdots p_N^{m_N}$. This tells us that to get *n* we must multiply p_1 by itself m_1 times and then multiply this number by p_2, m_2 times and keep going until we've multiplied m_N times by the number p_N. Notice the use of the notation $\cdots$ which we use for 'keep multiplying' (in contrast to . . . which we met before and which means keep going along some list). So in the example we've just seen where $n = 720$, we have $N = 3, p_1 = 2, p_2 = 3, p_3 = 5, m_1 = 4, m_2 = 2$ and $m_3 = 1$.

We'll now show that every natural number has a prime factorisation. We'll do this by using the 'proof by contradiction' technique that we employed in the first section to show that there is no largest natural number. This time we'll set our argument out in the way that a professional mathematician does it. In mathematics, new facts about numbers (or other mathematical objects) that are established through logical reasoning are called *theorems* and the arguments that we use to demonstrate these are called *proofs*. Theorems are given numbers that serve as labels to help us refer back to them. So the theorem that we are about to prove, which is that every natural number has a prime factorisation, will be called Theorem 1.2.1 where the first 1 refers to the chapter we are in, the 2 to the fact that we are in section two of that chapter and the second 1 tells us that this is the first theorem of that section. This result is so important that it is sometimes called the *fundamental theorem of arithmetic*.

Theorem 1.2.1. Every natural number has a prime factorisation.

Proof. Let n be the smallest natural number that doesn't have a prime factorisation. Clearly n cannot be prime and so it is composite. Write $n = bc$, then either b and c are both prime, or one of b or c is prime and the other is composite or both b and c are composite. We deal with each possibility in turn. Firstly suppose that b and c are both prime. Then $n = bc$ has a prime factorisation and we have our contradiction. If b is prime and c is composite we know that c is a smaller number than n and so it must have a prime factorisation. Now multiplying by the prime number b produces a prime factorisation for n and we again have a contradiction. The case where b is composite and c is prime works by the same argument. If b and c are both composite then each has a prime factorisation and multiplying them together gives us the prime factorisation for n that we need to establish the contradiction in the third and final case. $\square$

The symbol $\square$ is a convenient notation that mathematicians have developed to signal that the proof has ended.

We can make Theorem 1.2.1 into a sharper result by proving that not only does every natural number have a prime factorisation, but that this factorisation is *unique*, i.e. a number cannot have two different factorisations into primes. We will not prove this here (we are not going to prove everything in this book) but we will feel free to use this as a fact in future. You are invited to try to come up with a convincing argument yourself as to why this is true. Start with a natural number n and assume that it is the smallest one that has a prime factorisation using two different sets of prime numbers, and then see if you can get a contradiction. (Hint: You will need the fact, which we also haven't proved here, that if a prime number

p is a factor of a composite number n then it is also a factor of at least one of the divisors of n.)[4]

We've seen that prime numbers are defined to have no factors other than one and themselves. Later on in this book we will need to have some knowledge of *square-free numbers* and this is an ideal place to introduce them. These are precisely those natural numbers that have no factors that are squares (other than one). So 12 is not square-free as $12 = 3 \times 4 = 3 \times 2^2$ so 2^2 is a factor. All prime numbers are clearly square-free and so are numbers like $6 = 2 \times 3$. We start the list of square-free numbers as follows:

$$1, 2, 3, 5, 6, 7, 10, 11, 13, 14, 15, 17, 19, 21, 22, 23, 26, \ldots$$

Here are two useful facts about square-free numbers:

1. The square-free numbers are precisely those for which every prime number that occurs in the prime factorisation only appears once.

   So in the prime factorisation of a square-free number $n = p_1^{m_1} p_2^{m_2} \cdots p_N^{m_N}$ we must have $m_1 = m_2 = \cdots = m_N = 1$. For if any of these numbers is greater than 1 then we can pull-out a factor that is a square.

2. Every natural number can be written in the form

$$n = j^2 k,$$

where j and k are natural numbers and k is square-free.

   To see that this is true, first take $j = 1$, then $n = k$ and so we get all the square-free numbers. Now to get the rest of the natural numbers we need to be able to include all the squares and their multiples. To get the squares, just take $k = 1$ and consider all possible values of j. To get all numbers that are of the form $2j^2$ just take $k = 2$ and let j vary freely, and you should be able to work the rest out for yourself.

   For a numerical example, consider $n = 3120$. Then $j = 4$ and $k = 195$ since $3120 = 4^2 \times 195$ and $195 = 3 \times 5 \times 13$.

Our next task is to give Euclid's famous proof that there is no largest prime number. Before we do that, we need some preliminaries. Let n be a natural number. We use the notation $n!$ to stand for the product of all the numbers that start with 1 and finish with n, so $n! = 1 \times 2 \times 3 \times \cdots \times (n-1) \times n$. For example $1! = 1, 2! = 1 \times 2 = 2, 3! = 1 \times 2 \times 3 = 6, 4! = 1 \times 2 \times 3 \times 4 = 24$. I hope you spotted the useful formula $n! = n(n-1)!$, so that e.g. $5! = 5 \times 4! = 120$. $n!$ is pronounced 'n factorial' and it plays a useful role in applied mathematics and

---

[4] The standard proof can be found online at http://en.wikipedia.org/wiki/Fundamental_theorem_of_arithmetic

in probability theory as it is precisely the number of different ways in which n objects can be arranged in different orders.

The next thing we need are a couple of useful facts:

Fact 1. If a number b is divisible by a then it is divisible by every prime factor of a.

To see this, use Theorem 1.2.1 to write $a = p_1^{m_1} p_2^{m_2} \cdots p_N^{m_N}$. Now since b is divisible by a we can write $b = ac$ and so $b = p_1^{m_1} p_2^{m_2} \cdots p_N^{m_N} c$ and that's enough to give you the result we need.

We should also be aware of the 'logical negation' of Fact 1 which is that if b is not divisible by any prime factor of a then it is not divisible by a. This fact does not need a proof of its own as it follows from Fact 1 by pure logic.

Fact 2. If b_1 is larger than b_2 and a is a factor of both numbers then it is also a factor of $b_1 - b_2$.

Since a is a factor of both numbers we can write $b_1 = ac$ and $b_2 = ad$ and so $b_1 - b_2 = ac - ad = a(c - d)$, which does the trick. By the same argument you can show that a is a factor of $b_1 + b_2$ (whether or not b_1 is larger).

We're now ready to give the promised proof that there are infinitely many primes. This again appears in standard "theorem-proof" form and we apply the same proof by contradiction technique that we used before to show that there are infinitely many natural numbers.

Theorem 1.2.2. There are infinitely many prime numbers.

Proof. Suppose that the statement in the theorem is false and let p denote the largest prime number. Consider the number $P = p! + 1$. Our goal is to prove that there is a prime number q that is larger than p and this will then give us our contradiction. We'll show first that P is not divisible by any prime number on the list $2, 3, \ldots, p$. Let's start with 2. Suppose that P really is divisible by 2. Since $p! = 1 \times 2 \times \cdots \times p$ is divisible by 2, so is $P - p!$ by Fact 2 above. But $P - p! = 1$ which is not divisible by 2. So we have a contradiction and conclude that P is not divisible by 2. Now repeat the argument we've just given to see that P is not divisible by $3, 5, 7$ or any prime number up to and including p. Now if P is divisible by any composite number a, then by Fact 1 it must be divisible by every prime factor of a. We've already seen that P is not divisible by any prime number on our list and so we conclude that either P itself is prime (in which case $q = P$) or it is divisible by a larger prime than p (in which case q is larger than p but smaller than P). This gives the contradiction we were looking for and so we can conclude that there is no largest prime number. $\square$

Although there are an infinite number of prime numbers, these become rarer and rarer as we reach larger and larger numbers. The next result we'll prove tells us that somewhere within the vast multitude of natural numbers, we can find $m + 1$ numbers in succession, all of which are composite, for any m we can care to name.

Theorem 1.2.3. For any natural number m we can find $m + 1$ successive natural numbers, none of which is prime.

Proof. This is a 'proof by demonstration' where we simply show you how to construct what we need. Indeed I claim that the m successive composite numbers are given by the following list: $(m + 2)! + 2, (m + 2)! + 3, \ldots, (m + 2)! + (m + 1), (m + 2)! + (m + 2)$. I hope you agree with me that there really are $m + 1$ numbers on the list. Since $(m + 2)! = 1 \times 2 \times 3 \times \cdots \times m \times (m + 1) \times (m + 2)$, it is divisible by each of $2, 3, \ldots, m + 2$. It follows that $(m + 2)! + 2$ is divisible by 2 and so cannot be prime, $(m + 2)! + 3$ is divisible by 3 and cannot be prime, $\ldots$, $(m + 2)! + m + 2$ is divisible by $m + 2$ and so cannot be prime and that concludes what we aimed to show. □

Mathematicians like to play (this is how we discover new things) and it's fun to do this with the result we've just proved. Let's take $m = 2$. The theorem tells us that there are three successive numbers which are not prime and it even tells us where to find these – they are $4! + 2, 4! + 3, 4! + 4$ which are $26, 27, 28$. But by searching directly we can find much smaller numbers than this – indeed none of $8, 9$ and 10 are prime. Theorem 1.2.3 tells us that a certain list of m numbers exists but it doesn't give any information about the smallest number where such a list might begin. As far as I know, there is no known answer to that question. One of the reasons why prime numbers are so interesting is that it is easy to state unsolved problems. For example Goldbach's conjecture which dates back to 1742 remains unsolved. It says that every even number greater than 4 is the sum of two (odd) prime numbers. This is easy to verify for small numbers e.g. $6 = 3 + 3, 8 = 3 + 5, 10 = 3 + 7$ etc. but a general proof evades us so far.[5] By the way, the corresponding conjecture is false for odd numbers and you're invited to find a counter-example, i.e. an odd number which cannot be written as the sum of two prime numbers.

I've already commented on the fact that there is no known formula for generating all of the prime numbers. Mathematicians are fascinated with the patterns within prime numbers and the study of this comes within that part of the subject called *number theory*. On the other hand searching for very large prime numbers (and so discovering new ones) doesn't really involve much 'real' mathematics but does require a vast amount of computer power. The largest prime discovered so far (at the time of writing) was found in September 2008 and can be written as $2^{43112609} - 1$.[6] One of the quantities that mathematicians have

---

[5] See e.g. http://en.wikipedia.org/wiki/Goldbach's_conjecture for more background.
[6] See http://primes.utm.edu/largest.html

been particularly interested in is the 'function' $\pi(n)$ which is defined to be the number of primes less than or equal to the natural number n.[7] So for example, $\pi(2) = 1, \pi(3) = 2, \pi(4) = 2, \pi(5) = 3, \pi(10) = 4, \pi(100) = 25, \pi(1000) = 168, \pi(10^9) = 50847534$.[8] I emphasise that there is no exact formula that tells us what $\pi(n)$ might be for arbitrarily large n. However there is an approximate formula. In 1896, two French mathematicians Jacques Hadamard (1865–1963) and Charles de la Vallée-Poussin (1866–1962) independently published their proofs of what has now become known as the 'prime number theorem'. This tells us that as n becomes very large there is a sense in which $\pi(n)$ gets closer and closer to (but never reaches) the quantity $\frac{n}{\log_e(n)}$. Here $\log_e(n)$ is the *logarithm to base e of the number n*.[9] If you haven't met it before, you'll be able to find a lot of information about the number e in Chapter 7. Now let's focus on the phrase 'gets closer and closer to (but never reaches)'. My calculator tells me that $\frac{10}{\log_e(10)} = 4.34$ which isn't so far from the exact value of 4, $\frac{100}{\log_e(100)} = 21.71$ which is reasonably close to the precise value of 25, $\frac{1000}{\log_e(1000)} = 144.76$ which is in the right ball-park as 168 but $\frac{10^9}{\log_e(10^9)} = 48254942$ which isn't terribly close to 50847534. This doesn't look at all convincing but the evidence we have presented here is sparse – just a handful of numbers. Also I haven't yet told you the right way to compare $\pi(n)$ and $\frac{n}{\log_e(n)}$ – but maybe you can guess this? We won't prove the prime number theorem in this book as it uses far more advanced mathematics than we can hope to cover here, but one of the main themes we will present is the notion of a *limit* which gives a very precise meaning to this mysterious phrase 'gets closer and closer to (but never reaches)'. This should at least enable you to reach some understanding of what the prime number theorem is really telling us.

1.3 The Integers

The natural numbers are wonderful things but they have a number of limitations. One of these is that they only enable us to count in one direction. If for example a company wants to balance their assets against their debts then we really need to give the numbers a direction so that the positive benefits of the assets can be weighed against the negative impact of the debts. We do this by introducing

---

[7] The use of the Greek letter π here has absolutely nothing to do with the universal constant of that name which arises as the ratio of the circumference of any circle to it's diameter. Mathematics uses so many concepts that it's commonplace to employ the same symbol in different contexts where there is no fear of ambiguity leading to misunderstanding.

[8] See http://en.wikipedia.org/wiki/Prime_number_theorem

[9] If $a = b^x$ we say that x is the *logarithm to base b of the number a* and write $x = \log_b(a)$, e.g. $100 = 10^2$ and so $2 = \log_{10}(100)$.

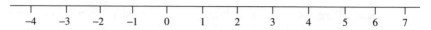

Figure 1.1. Integers as distances on a line.

another copy of the natural numbers and putting a minus sign in front of these. Thus we create the negative numbers $-1, -2, -3, -4, -5, \ldots$. We obtain the *integers* when these are combined together with the natural numbers. There is a neat way to see this visually as shown in Figure 1.1. Just draw a straight line on a piece of paper and mark the natural numbers in ascending order on the right so that there is a fixed distance between each successive number. We similarly mark the negative numbers in descending order on the left.

Observe that a new number has entered the arena which is neither positive nor negative. This is of course zero – denoted 0. If we think of the line as marking steps on a journey then 0 is the starting point, the natural numbers mark steps to the right away from zero and the negative numbers are steps to the left away from zero. The integers are then the numbers which describe all of these distances from zero on the line (in both directions and including standing still) and if we try to write them in increasing order we do something like this: $\ldots, -5, -4, -3, -2, -1, 0, 1, 2, 3, 4, 5, \ldots$ which indicates that these are infinite in both directions. Indeed we now have a geometric way of thinking about this twofold infinity in terms of our line which extends indefinitely in both directions – to the right and to the left. A little bit of (seemingly nit-picking) terminology will be useful for us later on. Natural numbers are sometimes also called *positive integers* and the natural numbers together with zero are called *nonnegative integers*.

The description that we have given of the integers so far is 'static'. We also need to be able to think of them 'dynamically', i.e. in a way that allows them to interact through arithmetic. Let's start by going back to Figure 1.1. Start at zero and take m steps to the right and then n steps to the left. We are then at the point $m - n$. If m is bigger than n we will be on the right and so have arrived at a natural number while if m is smaller than n we have reached a negative number. If m and n are equal then we are back at zero. Addition of integers corresponds precisely to combining steps in this way. So for example $7 + (-3) = 4$ is the result of taking 7 steps to the right and then 3 to the left. You can see similarly that $-7 + 3 = -4$. We have also seen that for any natural number n, $-n + n = 0 = n + -n = 0$.

Children often have difficulty in understanding the multiplication of integers. Now each non-zero number has a signature – it is either positive or negative and the rules for multiplication can be summed up in the following table:

$$\text{positive} \times \text{positive} = \text{positive},$$

$$\text{positive} \times \text{negative} = \text{negative},$$

$$\text{negative} \times \text{positive} = \text{negative},$$

$$\text{negative} \times \text{negative} = \text{positive},$$

so for example $-7 \times 4 = -28$ but $(-7) \times (-4) = -28 = 28$.[10] It's the last of these that often causes the most confusion. Why do two minuses make a plus? Here are two different attempts to explain this. One approach is algebraic and the other is dynamic and visual.

1. *Algebraic.* We have already seen that

$$n + -n = 0.$$

Now multiply both sides of this equation by -1 and you should agree that we get

$$-n + (- - n) = 0.$$

So we have that

$$n + (-n) = -n + (- - n).$$

Now add n to both sides (or cancel $-n$ from both sides) to conclude (after some rearrangement) that

$$n + (-n + n) = (- - n) + (-n + n).$$

But we now have

$$n + 0 = (- - n) + 0,$$

in other words $n = - - n$.[11]

2. *Dynamic.* Let's create the negative numbers from the natural ones. How shall we do this? We start with the number line as in Figure 1.1 but this time the negative numbers are missing as in Figure 1.2.

Now think of the line as though it were sitting inside of a two-dimensional infinite plane.[12] Draw a line that is perpendicular to our number line and which passes through zero. This is demonstrated in Figure 1.3.

Think of the perpendicular line as a mirror that creates an image of each natural number on the other side of the line. Then -1 is the mirror image of $+1$, -2 is the mirror image of $+2$ etc. -1 is a very special number in this context because it produces each reflected image by multiplication, i.e. $-1 \times 1 = -1, -1 \times 2 = -2, -1 \times 3 = -3$ etc. Now having gone from right to left, let's go back to the other side of the mirror by taking the mirror image again. To go from left to right we must multiply by -1 again so $-1 \times -1 = 1, -1 \times -2 = 2, -1 \times -3 = 3$ etc.

[10] The purpose of the brackets is to make the display easier to read.
[11] Some of you will notice our use of the commutative and associative laws of addition. I'm not going to make a fuss about these in this book. However they (and other algebraic properties of numbers) are listed in Appendix 4 for convenient reference.
[12] Mathematicians call this an 'embedding'.

11

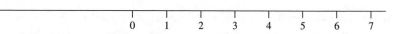

Figure 1.2. Natural numbers as distances along a line.

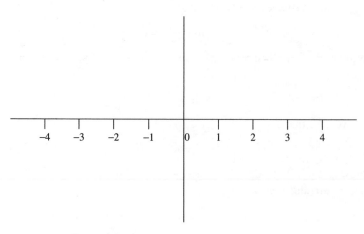

Figure 1.3. Obtaining the integers by reflection.

In this chapter we have worked entirely with whole numbers. Now we must split these apart and come to an understanding of how every point on the line in Figure 1.1 can represent a number. That is the task of the next chapter.

1.4 Exercises for Chapter 1

The first six questions in this section are designed to help you practise simple proof techniques and the remainder focus on prime numbers.

1. Prove that if n is an odd number then n^2 and n^3 are also odd. [Hint: Use the fact that if n is odd then $n = 2m - 1$ for $m = 1, 2, 3, \ldots$.]

2. Prove that if n^2 is an even number then n is also even. [Hint: Assume that n is odd and derive a contradiction.]

3. Prove that if m is odd and n is even then $m + n$ is also odd. If m is larger than n, is $m - n$ odd or even? Justify your answer with a proof.

4. Prove that the sum of four consecutive natural numbers is always even.

5. (a) Deduce that every odd number is either of the form $4m - 1$ or $4m + 1$ where m is a natural number. Numbers of the form $4m + 3$ are clearly also odd. How do you reconcile this with the previous assertion?
   (b) Show that the product of two odd numbers of the form $4m + 1$ is also of this form.
   (c) Show that the product of an odd number of the form $4m + 1$ and an odd number of the form $4m - 1$ is itself of the form $4m - 1$.

6. Show that any odd number that takes the form $3m + 1$ must also be of the form $6n + 1$, where m and n are natural numbers.

7. (a) It is rumoured that the numbers $2^p - 1$ where p is a prime number are always prime. Start making a list of these numbers and find the smallest p for which the rumour fails. The numbers of the form $2^p - 1$ which really are prime are called *Mersenne numbers* after a conjecture made by Marin Mersenne in 1644. How many of these can you find?
   (b) *Fermat numbers* (named after Pierre de Fermat) are those that are of the form $2^{2^n} + 1$ where n is a nonnegative integer ($n = 0, 1, 2, \ldots$). Deduce that the first five Fermat numbers are prime but that the sixth is composite. (Hint: Seek a factor between 640 and 645.)

8. Substitute $n = 1, 2, 3, \ldots$ into $n^2 - n + 41$. How many of the numbers that you obtain in this way are prime? Do you think that this formula always generates prime numbers? If not give a counter-example.

9. (a) Show that the proof of Theorem 1.2.2 still works if the number $p!$ is replaced by $\tilde{p}$ where p is the product of all the prime numbers up to and including p ($\tilde{p} = 2.3.5.7 \cdots p$).
   (b) Calculate the first six cases of $\tilde{p} + 1$. Show that the first five of these are all prime but that the sixth is composite. [Hint: Seek a factor between 55 and 65.]

10. Imitate the proof of Theorem 1.2.2 to show that there are infinitely many prime numbers of the form $4n - 1$. [Hint: Assume that there is a largest prime number p that takes this form and consider the number $M = 4(3.7.11 \cdots p) - 1$. Consider possible prime factors of the form $4m - 1$ and $4m + 1$ separately and use the results of Question 5.][13]

---

[13] There are also infinitely many prime numbers of the form $4n + 1$, but the proof of this fact is too difficult to include here.

2

Let's Get Real

Historically fractions owe their creation to the transition from counting to measuring.

Philosophy of Mathematics and Natural Science, Hermann Weyl

2.1 The Rational Numbers

A well-known nineteenth century mathematician named Leopold Kronecker (1823–1891) made a since oft-quoted remark at a meeting in Berlin in 1886 to the effect that 'The integers were made by God, everything else is the work of man'.[1] Whether or not we agree with this point of view, the needs of mathematics, science and technology require us to gain some insight into 'everything else'. We begin the (metaphorical) descent from heaven to earth by introducing the *rational numbers*. These are all numbers that can be written in the form $\frac{a}{b}$, where a is an integer and b is a natural number. So they comprise all (positive and negative) 'proper fractions' such as $\frac{1}{3}, -\frac{7}{9}, \frac{23}{201}$ as well as 'improper fractions' such as $-\frac{47}{13}$ and $\frac{1001}{19}$. We always write rational numbers in their lowest terms so e.g. $\frac{2}{4}, \frac{3}{6}, \frac{4}{8}$ etc. are all identified with $\frac{1}{2}$. Integers are included in the rational numbers – indeed we recognise the rational number $\frac{n}{1}$ as the integer n.

The name 'rational numbers' suggests that these are numbers that appeal to reason in some sense. This may well be the case but the terminology actually refers to the fact that such numbers express *ratios*, e.g. $\frac{3}{7}$ may express the division of a plot of land or a cake into 7 equal parts of which you or I may be entitled to precisely 3. We can do arithmetic with rational numbers and I hope that you remember how to add fractions by finding the lowest common multiple of the

---

[1] 'Die ganzen Zahlen hat der liebe Gott gemacht, ales andere ist Menschenwerk'.

denominators so e.g.

$$\frac{1}{2} + \frac{1}{5} = \frac{5}{10} + \frac{2}{10} = \frac{7}{10}.$$

The general rule for addition of fractions is

$$\frac{a}{b} + \frac{c}{d} = \frac{ad + bc}{bd},$$

but when you use this formula you should always cancel the right-hand side down to its lowest terms.

Similarly the general formula for multiplying fractions is

$$\frac{a}{b} \times \frac{c}{d} = \frac{ac}{bd}.$$

All rational numbers may be expressed in *decimal form* (or by a *decimal expansion*) so for example $\frac{1}{2} = 0.5$, $\frac{1}{4} = 0.25$, $\frac{32}{5} = 6.4$. The three numbers that I've written so far have finite decimal expansions but this is not always the case. Even the simple fraction $\frac{1}{3}$ can lead us to a contemplation of the infinite for its decimal expansion is $\frac{1}{3} = 0.3333333 \cdots$ where the dots indicate that there is no end to the process of writing 3s. This is sometimes written as $0.\dot{3}$ and called a *recurring decimal*. What does this infinite train of 3s actually mean? Well $0.3 = \frac{3}{10}$, $0.33 = \frac{3}{10} + \frac{3}{100}$ while $0.333 = \frac{3}{10} + \frac{3}{100} + \frac{3}{1000}$ so the meaning of $0.\dot{3}$ appears to be an infinite sum of fractions

$$0.\dot{3} = \frac{3}{10} + \frac{3}{100} + \frac{3}{1000} + \frac{3}{10^4} + \cdots + \frac{3}{10^{99}} + \cdots .$$

One of the tasks of later chapters will be to make sense of infinite sums of this type.

If I ask my pocket calculator to give me the value of $\frac{1}{3}$ it delivers the answer 0.3333333. This is a lie. Of course calculators are designed to help us make practical calculations and not to explore fundamental mathematical truths, so the approximation of the precise expression $\frac{1}{3}$ by the first eight digits of its decimal expansion 0.3333333 is probably going to be enough for most everyday purposes. There are many other fractions that fail to have a finite decimal expansion, e.g. $\frac{3}{11} = 0.27272727 \cdots$ which can be written more succinctly as $0.\dot{2}\dot{7}$.

An interesting fact about many fractions that have infinite decimal expansions is that they are *periodic* in the sense that there is a pattern that endlessly repeats itself. This is obvious in the case of $\frac{1}{3}$ or $\frac{3}{11}$. It is much less so if I ask my pocket calculator to find $\frac{1}{7}$. It tells me that the answer is 0.1428571 and there isn't any particular evidence of a pattern here, however if you continue to expand (using more powerful software or good old-fashioned long division) you obtain

$$\frac{1}{7} = 0.142857142857142857 \cdots ,$$

so the pattern here is an endless repetition of the digits 142857 first as $\frac{1}{10} + \frac{4}{100} + \frac{2}{1000} + \frac{8}{10^4} + \frac{5}{10^5} + \frac{7}{10^6}$ then as $\frac{1}{10^7} + \frac{4}{10^8} + \frac{2}{10^9} + \frac{8}{10^{10}} + \frac{5}{10^{11}} + \frac{7}{10^{12}}$ and so on, indefinitely. The 'dot notation' that was mentioned above for succinctly representing $\frac{1}{3}$ and $\frac{3}{11}$ in decimal form can also be extended to cases like this by placing a dot above both the first and last integers which appear in the block that is repeated, so $\frac{1}{7} = 0.\dot{1}4285\dot{7}$.

In fact it is possible to prove a theorem (although we won't do so here) to the effect that every fraction has a decimal expansion that is either

1. finite, e.g. $\frac{1}{4} = 0.25$,

2. periodic, as described above,

3. eventually periodic, i.e. periodic behaviour doesn't start immediately but it must do eventually (after a finite number of non-periodic numbers have appeared), e.g. $\frac{1}{6} = 0.166666\cdots = 0.1\dot{6}$.

Kronecker believed that rational numbers were 'less pure' than the integers, but we seem to commit an act of violence when we try to represent a fraction such as $\frac{1}{3}$ by an infinite decimal which will never have enough finite terms to capture the true meaning. There is another sense in which a decimal expansion is unsatisfactory and that is because it involves a 'choice of base'. When we write $\frac{1}{3}$ as a decimal, each successive term is obtained by taking a higher power of 10 on the denominator. But why should we privilege the number 10 in this way? In fact it is a convention that comes from the fact that standard human beings have ten fingers (or toes). If we used the number 2 as a base (as is common in binary arithmetic) we would write $\frac{1}{8} = 0.001$ because

$$\frac{1}{8} = \frac{0}{2} + \frac{0}{2^2} + \frac{1}{2^3},$$

indeed we will see later on in Section 6.11 that in this case the number $\frac{1}{3}$ has the recurring binary expansion $0.0101010\cdots = 0.\dot{0}\dot{1}$. But if we choose 3 as a base then we can avoid the infinite in dealing with $\frac{1}{3}$ as it has a finite 'trinary' expansion 0.1.

Now let us return to the number line on which we represented the integers in the last chapter as steps away from zero. The rational numbers give us much more flexibility as the diagram in Figure 2.1 shows.

Indeed it appears to the eye that the rationals are filling up all the spaces on the line. For example, let's try to get as close to zero as we can. We can take a very

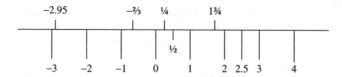

Figure 2.1. Some rational numbers on the number line.

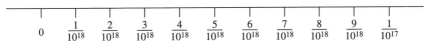

Figure 2.2. Magnified picture of the number line near zero.

small number indeed such as $\frac{1}{10^{17}}$. This seems imperceptibly close to zero and yet we can easily find nine smaller numbers $\frac{1}{10^{18}}, \frac{2}{10^{18}}, \ldots, \frac{9}{10^{18}}$ which are all smaller than $\frac{1}{10^{17}}$ as shown in Figure 2.2. But why stop at 17 and 18? By choosing larger and larger numbers in the denominator, we get smaller and smaller fractions that are surely filling up all the space on the line close to zero, aren't they? I often used to ask first year undergraduate mathematics students to vote on this question. About half of them usually agreed that the rational numbers really do fill up the whole number line. What do you think?

2.2 Irrational Numbers

To answer the question that we introduced at the end of the last section, we first need to focus on *square roots*. I'll remind you that a nonnegative rational number b has a square root a if $a^2 = b$. Let's list some numbers that have square roots. The square root of 0 is 0 as $0^2 = 0$. The number 1 has two square roots 1 and -1 as $1^2 = 1$ and $(-1)^2 = 1$. The number 4 also has two square roots 2 and -2. In fact if b is positive and it has a square root a that is positive then $-a$ is also a square root since $(-a)^2 = a^2 = b$. We always write the positive square root as $a = \sqrt{b}$ and so $-a = -\sqrt{b}$. So we've seen that $\sqrt{1} = 1$, $\sqrt{4} = 2$ and similarly $\sqrt{9} = 3$, $\sqrt{16} = 4$, $\sqrt{25} = 5$ etc. Some fractions also have easily obtainable square roots, e.g. $\sqrt{\frac{9}{16}} = \frac{3}{4}$. Now the numbers that we have looked at so far are all square roots of *perfect squares* (or fractions built from these) – in other words we have just identified the square root by noticing that the number we are square-rooting is $1^2, 2^2, 3^2, 4^2, \ldots$, But what about $\sqrt{2}, \sqrt{3}, \sqrt{5}, \sqrt{6}, \sqrt{7}$ etc.? Do these expressions have a meaning? We are now encountering what is sometimes referred to as the 'great crisis of Greek mathematics'. One of the great triumphs of antiquity is *Pythagoras' theorem* which states that in a right-angled triangle the length of the longest side c is related to that of the other two sides a and b by the beautiful formula:

$$c^2 = a^2 + b^2, \qquad (2.2.1)$$

so if, for example $a = 3$ and $b = 4$ then we must have $c = 5$. Now suppose that I put $a = 1$ and $b = 1$, then $c^2 = 2$ and I have constructed a right-angled triangle whose longest side $c = \sqrt{2}$ as is shown in Figure 2.3 below

17

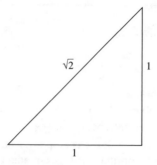

Figure 2.3. Right-angled triangle with sides of length 1, 1 and $\sqrt{2}$.

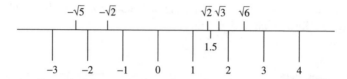

Figure 2.4. Irrational numbers on the number line.

Indeed if I look up $\sqrt{2}$ on my pocket calculator I get '$\sqrt{2} = 1.4142136$'[2] so I even know the first eight terms in the decimal expansion of this number. Now I can construct another right-angled triangle with smaller sides $a = \sqrt{2}$ and $b = 1$ to find that $c = \sqrt{3}$ and my calculator reports that '$\sqrt{3} = 1.7320508$'. Next I find $\sqrt{5}$ from a right-angled triangle whose shortest sides are of length 1 and 2. Since positive square roots are unique it follows that if three numbers x, y and z are related by $x = yz$, then $\sqrt{x} = \sqrt{y}\sqrt{z}$ and so we do not even need to draw a triangle to obtain $\sqrt{6}$ (although we can if we want to) – it must be $\sqrt{2} \times \sqrt{3}$. Now we have identified $\sqrt{2}$, $\sqrt{3}$, $\sqrt{5}$ etc. as lengths of the longest side of a right-angled triangle and we can even determine these to a high level of accuracy using our pocket calculators,[3] so we can incorporate these into the real number line as in Figure 2.4. Greek mathematicians knew about these things but they also knew the result of the next theorem and it is this which is the heart of the crisis alluded to earlier.

Theorem 2.2.1. If p is a prime number then $\sqrt{p}$ is not a rational number.

---

[2] The reason for using quotes is that (as we have already seen) calculators can't be trusted to tell the whole story when they deal with decimal expansions.

[3] Of course this presupposes the mathematics necessary to do these calculations.

Proof. Let us assume that $\sqrt{p}$ is rational and seek a contradiction. So we write $\sqrt{p} = \frac{a}{b}$ and then square both sides to get $p = \frac{a^2}{b^2}$ so that $b^2 = a^2 p$.[4]

Now we recall Theorem 1.2.1, that every natural number has a prime decomposition. We write a in terms of its prime decomposition as $a = 2^{m_1} 3^{m_2} 5^{m_3} \cdots q^{m_N}$ where q is the largest prime number that we need and N tells us how many prime numbers we have to count in order to get to q.[5] If we square this we get

$$a^2 = 2^{2m_1} 3^{2m_2} 5^{2m_3} \cdots q^{2m_N}.$$

Now we do the same for b. We write its prime factorisation as $b = 2^{n_1} 3^{n_2} 5^{n_3} \cdots r^{n_M}$ and square this to get

$$b^2 = 2^{2n_1} 3^{2n_2} 5^{2n_3} \cdots r^{2n_M}.$$

Now let's return to the equation $b^2 = a^2 p$ and substitute in our prime factorisations for a^2 and b^2. We obtain

$$2^{2n_1} 3^{2n_2} 5^{2n_3} \cdots r^{2n_M} = 2^{2m_1} 3^{2m_2} 5^{2m_3} \cdots q^{2m_N} p.$$

Now if the number p doesn't appear on the left-hand side we already have a contradiction so let's assume that it does and that it is one of the numbers $2, 3, 5, \ldots, r$. Each of these prime numbers appears an *even* number of times on the left-hand side. Now on the right-hand side either p is not one of the numbers $2, 3, \ldots, q$ in which case it only appears *once* altogether, or it is in that list of numbers in which case the extra multiplication by p means that it appears an *odd* number of times. Either way we have a contradiction and so we conclude that $\sqrt{p}$ cannot be a rational number. $\square$

The conventional story is that Greek mathematicians BCE[6] were not prepared to accept that quantities that could not be expressed as rational numbers could be legitimate numbers. It wasn't until the Renaissance that mathematicians began to feel comfortable with these numbers and this is believed to have held back the development of science and mathematics for several centuries. Nowadays we call numbers like $\sqrt{p}$ *irrational numbers*. This is not because we believe their existence to be an affront to reason – the word 'irrational' here should be interpreted as 'not a ratio of integers'.

In mathematics we use the term 'corollary' to label a new result which follows fairly easily from a theorem that has just been proved without us having to do very much extra work. The first corollary in this book is one that we obtain from Theorem 2.2.1:

---

[4] Of course you can assume that the fraction is written in its lowest terms but this isn't necessary for the proof to work.

[5] If for example, 3 doesn't appear in the prime factorisation of a, the notation we've used still makes good sense as we then have $m_2 = 0$ and so $3^{m_2} = 3^0 = 1$.

[6] BCE is a secular term meaning "before the common era" and is used as an alternative to BC ("before Christ") – see e.g. http://en.wikipedia.org/wiki/Common_Era

Corollary 2.2.1. There are an infinite number of irrational numbers.

Proof. We saw in Theorem 1.2.2 that there are an infinite number of prime numbers while Theorem 2.2.1 tells us that each of these prime numbers has an irrational square root. It follows that the list $\sqrt{2}, \sqrt{3}, \sqrt{5}, \sqrt{7}, \ldots$ has no largest member and that's what we set out to prove. □

We have shown that there are already infinitely many irrational numbers that take the form $\sqrt{p}$ where p is a prime number. In fact there are many more irrational numbers than this and later on in, Chapter 7, we will look at proving the irrationality of some very famous numbers such as e – the base of natural logarithms and π – the ratio of the circumference of any circle to its diameter. But for now let's stay with square roots for just a little longer. We've seen that all prime numbers have irrational square roots and we know that the perfect squares have square roots that are integers. What about the square roots of other composite numbers such as 6, 8, 10, 12, 14, 15 etc.? The answer to this question can be found in the following theorem:

Theorem 2.2.2. If N is a natural number then either it is a perfect square or $\sqrt{N}$ is irrational.

Proof. Suppose that N is not a perfect square and suppose that $\sqrt{N}$ is a rational number. We'll try to obtain a contradiction. First we write $\sqrt{N}$ as a (nonnegative) integer plus a fraction in its lowest terms i.e.

$$\sqrt{N} = a + \frac{b}{c}, \tag{2.2.2}$$

where a, b and c are natural numbers. We also know that b is a smaller number than c with the fraction $\frac{b}{c}$ being written in its lowest terms. Now multiply both sides of the expression labelled (2.2.2) by c and square both sides to get

$$c^2 N = (ca + b)^2$$
$$= c^2 a^2 + 2cab + b^2.$$

Using some simple algebraic manipulations we see that

$$b^2 = c^2 N - c^2 a^2 - 2cab$$
$$= c(cN - ca^2 - 2ab).$$

This tells us that c is a factor of b^2 so that $b^2 = cd$ where $d = cN - ca^2 - 2ab$ is a natural number. So $c = \frac{b^2}{d}$. If you substitute for c in (2.2.2) you can check that you get

$$\sqrt{N} = a + \frac{d}{b}. \tag{2.2.3}$$

Now we've already pointed out that b is smaller than c. We also have that d is smaller than b for otherwise cd would be the product of two numbers both greater than b and this must give a larger number than b^2.[7] That would contradict the fact that $b^2 = cd$. This means that $\frac{d}{b}$ is a fraction written in lower terms than $\frac{b}{c}$ and that gives the contradiction we need to prove the theorem. □

So Theorem 2.2.2 tells us that $\sqrt{6}$, $\sqrt{8}$, $\sqrt{10}$ etc. are all irrational numbers. By using some simple algebra we can construct many more irrational numbers using those that we already have. I could present the following in standard 'theorem-proof' mode but let's have a change and use a list. (The label 'I' here stands for 'irrational'.)

Algebra of Irrational Numbers

I(i) *If x is an irrational number and a is rational and non-zero then ax is irrational.*

To see that this is true suppose that ax is rational and look for a contradiction in the usual way. So we write $ax = \frac{q}{p}$ where q is an integer and p is a natural number. But a is rational so $a = \frac{c}{d}$. This tells us that

$$\frac{c}{d}x = \frac{q}{p},$$

i.e. $x = \frac{dq}{cp}$ which is rational and that's the contradiction we were looking for.

So we now see that numbers like $12\sqrt{3}$ and $\frac{\sqrt{5}}{2}$ are irrational.

I(ii) *If x is irrational and a is rational then $x + a$ is irrational.*

This is more-or-less the same argument as the one we just used so we suppose that $x + a$ is rational and write $x + a = \frac{e}{f}$. But $a = \frac{b}{c}$ is rational and so

$$x = \frac{e}{f} - \frac{b}{c} = \frac{ec - bf}{cf},$$

which is a rational number and that's the contradiction we wanted.

This result together with that of Theorem 2.2.2 tells us that numbers like $100 + \sqrt{99}$ are irrational. If we also combine it together with that of (I(i)) we see that the number $\frac{1}{2} + \frac{\sqrt{5}}{2}$ is also irrational. This is a very famous number – it was called the *golden ratio* (or *golden section*) by the Ancient Greeks. We will have more to say about it later on in the book.[8]

I(iii) *If x is a positive irrational number then $\sqrt{x}$ is also irrational.*

To see this we suppose that $\sqrt{x} = \frac{g}{h}$ is rational, but then $x = (\sqrt{x})^2 = \frac{g^2}{h^2}$ is also irrational and there is the contradiction we needed.

---

[7] d cannot be equal to b. Why?
[8] If you can't wait then have a look at http://en.wikipedia.org/wiki/Golden_ratio

As an example let's look at $\sqrt{2} + \sqrt{3}$. We'll show it is irrational by using I(iii) to express it as the square root of an irrational number, indeed

$$(\sqrt{2} + \sqrt{3})^2 = 2 + 3 + 2\sqrt{2}\sqrt{3} = 5 + 2\sqrt{6},$$

which is irrational by I(i) and I(ii). You might try to generalise this argument to show that $\sqrt{p} + \sqrt{q}$ is irrational whenever p and q are distinct prime numbers. However, beware – it is not true that the sum of any two irrational numbers is itself irrational, e.g. $\sqrt{5} + (4 - \sqrt{5}) = 4$.

I(iv) *If a is an irrational number then so is $\frac{1}{a}$.*
The argument that proves this is a fairly simple variation on those given above and so I feel comfortable leaving it to you to do yourself.

All irrational numbers have decimal expansions but I'm not going to tell you how to find them at this stage. Here are some examples where I've given the first nine decimal places:

$$\sqrt{2} = 1.414213562\ldots$$

$$\sqrt{3} = 1.732050808\ldots,$$

The golden section $\dfrac{1}{2} + \dfrac{\sqrt{5}}{2} = 1.618033988\ldots,$

$$\pi = 3.141592654\ldots.$$

Decimal expansions of irrational numbers are always infinite and can never be periodic or even eventually periodic. The same is true if we choose expansion in any base other than 10. This means that irrational numbers are less tangible than rational numbers. If we only know a rational number through its decimal expansion then we know that eventually we will find some periodic pattern (though we may have to go beyond (say) a googol of decimal places to find this). On the other hand, the decimal expansion of an irrational number is essentially unknowable, although amateur mathematicians have lots of fun in calculating numbers like π to greater and greater precision. Indeed as of August 2010 it was known to 5,000,000,000,000 decimal places (see e.g. http://en.wikipedia.org/wiki/Chronology_of_computation_of_%CF%80). On the other hand let's suppose that you are playing a game with a friend wherein they give you a number and you have to guess what it is. Let us suppose that they give you the first 5,000,000,000,000 decimal places of the number and that you are clever enough to recognise these as coming from π. Can you then say that the number really is π? The answer is no! For all you know you might have been presented with a rational number that agrees with π to the first 5,000,000,000,000 decimal places but is (for example) zero from the 5,000,000,000,001st decimal place onwards.

2.3 The Real Numbers

Let's return to the number line drawn in Figure 1.1. We tried to fill this line up with rational numbers representing every distance from the origin to a point on the line. This process failed. We have seen that there are infinitely many irrational numbers like $\sqrt{2}$ that are not rational but which still measure legitimate distances along the line. When we combine the rational and irrational numbers together there are no more gaps. These numbers are enough to measure every distance along the line. A number that is either rational or irrational is called a *real number*. Note that (as usual) the word 'real' here is just a name – you should not believe that real numbers are any more or less 'real' than other types of number that you may encounter.[9] At this stage you should have a reasonable intuition as to what a real number is, but be aware that we haven't really given a satisfactory definition of them. This is because we haven't properly defined irrational numbers – all we've done is provide some examples of them. We could give a working definition of a real number as 'a number that has a decimal expansion' but that is far too indirect as is the geometric definition in terms of distances along a line from an arbitrary point. Real numbers lie at the heart of much of pure mathematics and are central to applications in science and engineering, and yet they are strangely elusive. We will return to the problem of how they should be properly defined – but not until the very end of the book.

Not knowing what real numbers are shouldn't stop us working with them and so we'll take for granted that we can add, subtract, multiply and divide real numbers and that the answer is always a real number (as long as we don't divide by zero). How do we add two real numbers? Just add the decimal expansions in the usual way. Of course you will never see the whole number but the first few terms is enough isn't it? For example $\sqrt{2} + \pi = 4.555806216\ldots$. We will always assume in this book that real numbers obey the standard algebra that we expect of numbers. So for example we assume that the commutative law of addition holds so that for all real numbers a and b:

$$a + b = b + a,$$

and we will assume that multiplication and addition interact through the distributive law,[10] i.e.

$$a(b + c) = ab + ac,$$

for all real numbers a, b and c. It is fairly easy algebra to prove these for rational numbers. For real numbers this must await a proper definition.

---

[9] For example 'imaginary', and more generally 'complex numbers' but we won't consider these until Chapter 8, and even then they only make a brief appearance.

[10] See Appendix 4 for a full list of these algebraic properties of numbers.

Before we (temporarily) leave the realm of the finite, it's worth listing some fascinating facts about irrational numbers which demonstrate their relative 'weight' in comparison to the rational numbers.

- There are infinitely many rational numbers and infinitely many irrational numbers but there are more irrationals than rationals! The irrational numbers lay claim to a higher order of infinity.
- There are an infinite number of irrational numbers between every two rational numbers p and q – no matter how close together p and q might be on the real number line.
- There is a rational number between every two irrational numbers.
- Consider all the real numbers between 0 and 1. This forms a portion of the number line that has length 1. Take away all the fractions on that line. Then the length of the line is still 1 – so the rational numbers contribute nothing to the length of the line.

We'll give proofs of all of these facts later on (except the last which is a little too sophisticated for this book). At the moment, you might be thinking that the real number line is a much more complicated object than you expected – and that is a very good way to be thinking.

I'll make one more comment on this for now. One way of dividing up the world is into the 'discrete' and the 'continuous'. Discrete phenomena are separated from each other like cows in a field or the beats of a drum. Continuous phenomena appear to flow into each other like the paint on the wall next to my computer. All of our direct experience with numbers is with the discrete through counting (the natural numbers) or dividing into pieces (the rationals). But the real numbers are at a different level as they measure the 'continuum' that is represented by the number line. The essential difference between the discrete and the continuous is captured by the irrationals and this is one of the reasons why we find them so strange. They are the first type of number to take us away from our direct experience of the world around us.

2.4 A First Look at Infinity

The real numbers are all finite numbers. No matter how difficult it may be to pin down an irrational number through its decimal expansion, it still represents a fixed point on the number line that measures a finite distance from the origin. But the line is infinite in extent. What does this really mean? Well this is the same phenomenon that we encountered at the end of Section 1.1. No matter how large a real number we take, we can always extend the line a little bit further in length to get larger real numbers. But what about the full infinite extent of the line? Can this be represented by a number?

We are always taught at school that division by zero is not allowed but let's define a new 'number' which we will call 'infinity' and represent by the symbol ∞ as follows:

$$\infty = \frac{1}{0}. \tag{2.4.4}$$

Now whatever it is ∞ cannot be a real number, but there is some sense to (2.4.4) for, dividing by a very small number always gives a very large number (e.g. $\frac{1}{10^{-26}} = 10^{26}$) – so why not go the whole hog and divide by zero? Well we can (up to a point) but as we will see – we have to be very careful.

If you read Section 1.1 again, your first question might be – what about $\infty + 1$? Surely this must be a bigger number than ∞? Let's find out by adding in the usual way and then

$$\infty + 1 = \frac{1}{0} + 1 = \frac{1 + (1 \times 0)}{0} = \frac{1 + 0}{0} = \frac{1}{0} = \infty.$$

So $\infty + 1 = \infty$, which is intuitively neat as it tells us that once we get to infinity we can't get any bigger by adding one – the mighty infinite absorbs the puny finite into itself. Indeed by the same argument that we just presented you can show that $\infty + c = \infty$ for any real number c.

We also have

$$\infty + \infty = 2\infty = 2\frac{1}{0} = \frac{1}{\frac{0}{2}} = \frac{1}{0} = \infty,$$

and by a similar argument $c \times \infty = \infty$ for any real number c.

But before we get carried away let's return to $\infty + 1 = \infty$. Surely if (as we've assumed so far) normal arithmetic applies, then we should be able to subtract ∞ from both sides of this equation to get $1 = \infty - \infty$. But we also have $\infty + 2 = \infty$ and that yields $2 = \infty - \infty$, so we have found that $1 = 2$ which is nonsense. Now at this point we can stop and conclude that defining $\infty = \frac{1}{0}$ is a bad definition as it leads to a contradiction, or we can continue on the basis that the rules of ordinary arithmetic are suspended for infinite numbers and that, in particular, $\infty - \infty$ makes no sense and so is banned from mathematics. Let's continue on that basis and see what happens. Since e.g. $1 \times \infty = 2 \times \infty = \infty$ we can also reject $\frac{\infty}{\infty} = \frac{0}{0}$ as being meaningless expressions.

Now if you've accepted the story so far you might argue that there should be two infinite numbers – one that measures the positive infinite distance from zero to the never-ending right of the number line and another that measures the distance to the left. If we identify ∞ with the infinite positive quantity, then the negative one should be $-\infty = \frac{-1}{0}$. But notice that

$$\frac{-1}{0} = \frac{-1 \times 1}{0} = \frac{1}{\frac{0}{-1}} = \frac{1}{0} = \infty,$$

25

2 LET'S GET REAL

so we've shown that $-\infty = \infty$!^{11} Now either this tells us that infinite numbers are too crazy for their own good, or perhaps that the infinite line should really be extended to some sort of infinite circle where positive and negative infinities can meet and merge. In any case it's time to stop this limited and unsatisfactory exploration of the infinite. From now on, any attempted use of '$\infty = \frac{1}{0}$' will be banned from our mathematical world. In Chapter 10 we'll meet some of the ideas of the nineteenth century German mathematician Georg Cantor who gave a more sophisticated approach to infinite numbers. For now, it's time to get our feet back on the ground!

2.5 Exercises for Chapter 2

1. Exhibit the number 0.8125 in binary notation.

2. Use long division to write $\frac{1}{13}$ as a recurring decimal.

3. (a) Consider the recurring decimal $x = 3\dot{9}$. By writing $99x = 100x - x$, show that $x = \frac{13}{33}$ as a fraction in its lowest terms.
   (b) Use a similar technique to write $0.510\dot{7}$ as a fraction in its lowest terms.

4. Write down five rational numbers between $5\frac{10}{61}$ and $5\frac{1}{6}$.

5. Which (if any) of the following numbers are irrational (a) $\sqrt{1296}$, (b) $\sqrt{1297}$? (Try to answer this question without using a calculator.)

6. Is it true that if x and y are irrational numbers then their product xy is always irrational? Give a careful proof or a counter-example to the claim.

7. This question develops an alternative proof (which is very often presented in textbooks) that $\sqrt{2}$ is irrational. Assume $\sqrt{2}$ is rational and so can be written $\frac{p}{q}$ as a fraction in its lowest terms. Square both sides and so deduce that p^2, and hence p, is even. Now write $p = 2m$ for some natural number m. Show that q is even and find the contradiction in our assumption.

8. Use a similar argument to that of Question 7 to show that $\sqrt[3]{2}$ is irrational.

9. Can you find two irrational numbers a and b such that a^b is rational? [Hint: Thinking about the number $\sqrt{2}^{\sqrt{2}}$ is a good place to start.]

10. In 1933, a Cambridge undergraduate, David Champernowne, studied the real number 0.1234567891011121314151617181920 21 ... formed by

---

^{11} The identification of ∞ with $-\infty$ has a sound geometric intuition behind it – see the section on *inversion in the circle* in 'To Infinity and Beyond' by Eli Maior. This book is briefly discussed in the Further Reading section.

the natural numbers written in sequence. Is this number rational or irrational? Give a reason for your answer. Write down one rational and one irrational number strictly between Champernowne's number and $0.\dot{1}2345678910$.

11. Assume that ∞ can be treated like an ordinary number. What sense can you give to $\sqrt{\infty}$?

3

The Joy of Inequality

"Are you content now?" said the Caterpillar. "Well I should like to be a little larger, Sir, if you wouldn't mind," said Alice

Alice in Wonderland, Lewis Carroll

3.1 Greater or Less?

When we first learn mathematics, equations are a dominant theme. The essence of an equation lies in the fact that there is an 'equals sign' $=$ which separates two seemingly distinct expressions, and our job is to solve the equation which involves finding conditions under which the two expressions really are the same. For example consider the quadratic equation

$$3 - x^2 = 9 - 5x. \tag{3.1.1}$$

With a little bit of algebra we can show that this is equivalent to

$$x^2 - 5x + 6 = 0,$$

which has two solutions $x = 2$ and $x = 3$. So the expression on the left-hand side of (3.1.1) is equal to that on the right-hand side when $x = 2$ or when $x = 3$. Otherwise they will be unequal.

In this chapter we will shift the emphasis from equality to inequality. Of course we are much more used to seeing inequality than equality in the world around us, so it should be no surprise that mathematicians have developed extensive techniques for investigating this. We will see that these techniques are absolutely vital for exploring the world of limits.

28

Let's try to compare two distinct real numbers a and b. The most basic inequality of all is

$$a \neq b, \qquad (3.1.2)$$

where the line through $=$ nicely symbolises the act of cancelling the possibility that a and b might be equal. So (3.1.2) would be satisfied if e.g. $a = 2.713$ and $b = 2.714$. This is all very nice but it won't take us very far. The serious players in the study of the unequal are four rather more subtle relations which we denote $<, \leq, >$ and $\geq$. Let's meet these one at a time. We write $a < b$ whenever a is strictly smaller than b. The word 'strictly' used here emphasises the fact that equality $a = b$ is excluded. Equivalently $a < b$ whenever $b - a$ is a positive real number. So for example $3.12 < 3.1213$. We say that $a \leq b$ whenever a is either smaller than or equal to b and the line under the $<$ sign indicates that $a = b$ is no longer excluded. So we can certainly write $3.12 \leq 3.1213$ but it is also correct to say $3.12 \leq 3.12$. Having the two symbols $<$ and $\leq$ may seem like hair-splitting but we will see that the distinction that these allow us to be able to make can be quite crucial. Now if a is less than b then b is greater than a and the symbol $>$ nicely encapsulates this reversal of roles, indeed we can define $a > b$ to mean the same thing as $b < a$ and $a \geq b$ to be equivalent to $b \geq a$. You might like to try proving the (obvious?) fact that $a \leq b$ and $b \leq a$ if and only if $a = b$. It's also worth pointing out that $a < b$ implies $a \leq b$, but the converse statement is false.

The symbols $<, \leq, >$ and $\geq$ are sometimes called *order relations* as they allow us to capture the natural order structure of the real number line whereby numbers get greater as you move to the right. Be aware though that when we consider negative numbers we have for example $-2 < -1$ and sometimes people find this to be counter-intuitive as -2 has a larger magnitude than -1. But -2 is further to the left than -1 on the real number line and so is smaller, as shown in Figure 3.1. If in doubt, it's always best to go back to the definition. Recall that we've agreed that $a < b$ means that $b - a$ is positive. Now put $a = -2$ and $b = -1$ then $b - a = -1 - (-2) = -1 + 2 = 1$ which is certainly a positive number.

We need to be able to manipulate inequalities and understand how they interact with addition, substraction, multiplication and division. We already have one important rule which is effectively the definition of $<$, but it's worth writing this down again and giving it a name (L1) so that we can refer back to it later (the L here stands for 'less than' as we are collecting together a list of rules for manipulating the 'less than' symbol $<$):

(L1) $a < b$ if and only if $b - a > 0$.

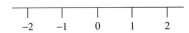

Figure 3.1. $-2 < -1 < 0 < 1 < 2$.

We also have a version of this for $\leq$:

(L1)′ $a \leq b$ if and only if $b - a \geq 0$.

I'll now list and prove some other very useful rules for manipulating inequalities. All of these will be expressed in terms of $<$ but they all have $\leq$ versions which are obtained by carefully substituting every $<$ with $\leq$ in a systematic manner.

3.1.1 Algebra of Inequalities

(L2) Adding a constant. If $a < b$ then

$$a + c < b + c \text{ for any real number } c.$$

To prove this we'll use a nice little technique which I call the 'mathematician's favourite trick' (or MFT for short). So I'll tell you what MFT is first before we go further. Suppose that we have an expression (which might be quite complicated) that is equal to some real number x. First write

$$x = x + 0,$$

then we can write $0 = y - y = -y + y$ where y is any real number, so that

$$x = x + (-y + y) = (x - y) + y.$$

Now the point of the MFT is that in some situations we can find a y so that both $x - y$ and y are easier to deal with than x and this is often the key to making further progress. Don't worry if this seems rather strange to you – we'll be using the MFT quite a lot in this book so it should be quite a familiar tool by the time we reach the end.

Now we'll prove (L2):

If $a < b$ then $b - a > 0$ so by MFT

$$b + c - a - c > 0,$$

$$\text{i.e. } (b + c) - (a + c) > 0,$$

$$\text{i.e. } a + c < b + c.$$

You can extend (L2) to a stronger result by more or less the same reasoning, to show that if a, b, c and d are four real numbers such that $a < b$ and $c < d$ then $a + c < b + d$. You may like to try to prove that yourself. I hope I won't confuse anyone if I also refer to that result as (L2) later in the book.

(L3) Multiplying by a positive number. If $a < b$ and $c > 0$ then

$$ac < bc.$$

I won't prove (L3) as it's quite easy.

(L4) Multiplying by a negative number. If $a < b$ and $c < 0$ then

$$ac > bc.$$

Before we prove (L4) let's discuss it. First of all it's a little less obvious than (L3) as it tells us that multiplying by a negative number changes the direction of an inequality. If you think this is strange, it's always best to do a few numerical experiments for yourself before going further. So let's take $a = -3$ and $b = -2$. We know $-3 < -2$ so $a < b$ is satisfied in this case. Now choose $c = -2$. Then $ac = -2 \times -3 = 6$ and $bc = -2 \times -2 = 4$. Of course $6 > 4$ so $ac > bc$ as was promised.

To prove (L4) we use (L1) and notice that $b - a > 0$ so it is a positive number. We've required c to be a negative number and we know that a positive number multiplied by a negative number is always negative (see Section 1.3). So $c(b - a) < 0$ (as it is negative). Now multiply out the bracket to get $bc - ac < 0$ and that tells us that $ac > bc$ which is what we wanted.

(L4) takes a very special form when $c = -1$. In this case it tells us that if $a < b$ then $-b < -a$ and we'll use this quite often. You can see this as a consequence of the way that multiplication by -1 acts as a reflection through a mirror as we discussed at the end of Chapter 1 and as shown in Figure 3.2.

We've just seen that multiplication by negative numbers has the effect of reversing inequalities. Another way of doing this is by 'inversion'.

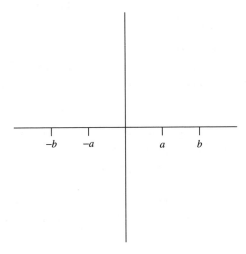

Figure 3.2. If $a < b$ then $-b < -a$.

(L5) Inversion. If a and b have the same sign (i.e. both are positive or both are negative) and $a < b$ then

$$\frac{1}{b} < \frac{1}{a}.$$

Again before you prove this – try an experiment if it seems strange, e.g. check what happens when $a = 2$ and $b = 4$.

To prove (L5) again use (L1) to transform $a < b$ to $b - a > 0$. Now since a and b both have the same sign we have $ab > 0$. Now take $c = \frac{1}{ab}$ in (L3) and we have that $\frac{b-a}{ab} > 0$. The result we are seeking follows when we do a little algebra and observe that $\frac{1}{a} - \frac{1}{b} = \frac{b-a}{ab} > 0$.

Be aware that (L5) fails if a and b have opposite signs e.g. $-2 < 3$ but $\frac{1}{3} > -\frac{1}{2}$. You should be able to conjecture and then prove the replacement for (L5) in this case.

(L6) Squaring. If a and b are both nonnegative then

$$a \leq b \quad \text{if and only if}^1 \quad a^2 \leq b^2.$$

To prove this we use the well-known algebraic identity

$$b^2 - a^2 = (b - a)(b + a),$$

from which we see that the left-hand side is positive (or negative, respectively) if and only if the right-hand side is and since $b + a \geq 0$ the sign of $b^2 - a^2$ is the same as the sign of $b - a$. You should be able to see the rest from there.

We have already pointed out that (L2) to (L6) have straightforward extensions to the case where $<$ is replaced by $\leq$. They can also all be adapted to the cases where we have $>$ and $\geq$ (indeed these follow immediately by using the fact that $>$ is defined in terms of $<$ and $\geq$ in terms of $\leq$) so that e.g. (L4) becomes

$$\text{If } a > b \text{ and } c < 0 \text{ then } ac < bc.$$

Here's a beautiful application of inequalities. We'll show that between any two rational numbers (no matter how close they are) there are an infinite number of irrational numbers – so e.g. there are infinitely many irrational numbers between 0.49999999 and 0.5.

Theorem 3.1.1. Given any two rational numbers a and b with $a < b$ we can find infinitely many irrational numbers q such that

$$a < q < b.$$

Proof. Let p be a prime number. Since $p > 1$ then $\sqrt{p} > 1$ by (L6) and so $\frac{1}{\sqrt{p}} < 1$ by (L5). Now define $q = a + \frac{1}{\sqrt{p}}(b - a)$. q is irrational by I(i) and I(ii).

---

[1] If p and q are propositions then 'p if and only if q' means that p implies q and q implies p.

$$0 \qquad a \qquad b$$

Figure 3.3. The interval between a and b is shaded.[4]

Since there are infinitely many prime numbers it follows that there are infinitely many numbers of this form. We will prove that $a < q < p$. Now $q > a$ since $q - a = \frac{1}{\sqrt{p}}(b - a) > 0$ and $q < b$ since

$$b - q = b - a - \frac{1}{\sqrt{p}}(b - a) = (b - a)\left(1 - \frac{1}{\sqrt{p}}\right) > 0$$

and the result is established. □

3.2 Intervals

Intervals are very important parts of the real number line. They work like this. Fix a and b so that $a < b$. An interval (see Figure 3.3) is the collection[2] of all numbers that lie between a and b. That's a bit vague as we haven't said anything about whether the end-points a and b should be included. If we worry about these we get four different intervals that extend from a to b.

(a, b) is called an *open interval*. It comprises all the points lying strictly between a and b but the end-points a and b themselves are NOT included.[3]

$[a, b)$ and $(a, b]$ are called *half-open intervals*. The first of these includes all of the points in (a, b) together with a but not b while the second has b in it but a is excluded.

$[a, b]$ is called a *closed interval*. It contains all of the points in (a, b) as well as both end-points a and b.

Note that, for example, 1 is in the interval $(-2, 1]$ but it is not in $[-2, 1)$. However 0 is in both of these intervals. We can also express intervals quite nicely using inequalities (which is one reason why I included them in this chapter). So a number x is in the interval (a, b) if and only if $x > a$ and $x < b$. Whenever we have two inequalities like this it is notationally convenient to combine them and write $a < x < b$ which translates precisely as 'x is larger than a but smaller than b' (or 'x is *between* a and b') and this is precisely what it means to be inside the

---

[2] Technically speaking we should say 'set' (see Appendix 2).

[3] Some mathematicians prefer to use the notation $]a, b[$ for open intervals and $[a, b[$ and $]a, b]$ for the half-open intervals $[a, b)$ and $(a, b]$, respectively.

[4] I deliberately haven't said if the interval is open or closed as you can't easily show this on a diagram.

interval (a, b). We won't do anything more with intervals for now – but we'll meet them again later on.

3.3 The Modulus of a Number

When we're dealing with numbers there are occasions when the sign of the number, i.e. whether it is positive or negative, is absolutely crucial, and other times when all that matters is the magnitude or size of the number. To deal with the latter case we introduce a neat notation $|x|$ and we call this the *modulus* of the real number x. So e.g. $|7| = 7$ but $|-23| = -(-23) = 23$. A slick way of defining the modulus in terms of $-$, $\geq$ and $<$ is

$$|x| = \left\{ \begin{array}{ll} x & \text{if } x \geq 0 \\ -x & \text{if } x < 0, \end{array} \right.$$

so that in particular $|0| = 0$. Notice that the definition takes advantage of the fact that if x is negative, then $x = -|x|$.

It's worth pointing out that we get two very obvious (but sometimes useful) inequalities from this definition:

$$x \leq |x| \text{ and } -x \leq |x| \tag{3.3.3}$$

Figure 3.4 shows a graph of $|x|$ plotted against x.

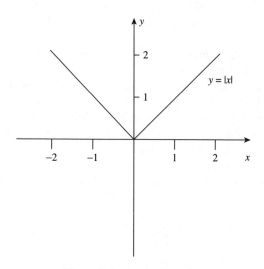

Figure 3.4. Graph of $y = |x|$.

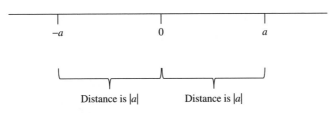

Figure 3.5. The modulus as a length.

The way in which the modulus interacts with the $+$ operation gives rise to one of the most important inequalities in the whole of mathematics. It is called the *triangle inequality* and we'll present it as a theorem.

Theorem 3.3.1. For all real numbers a and b,

$$|a + b| \leq |a| + |b|. \tag{3.3.4}$$

Now before we prove this theorem, let's make a few remarks. The first question you might ask is – why is it $\leq$ here? Why not $=$? After all if you put $a = 2$ and $b = 1$ the left-hand side and right-hand side of (3.3.4) are both 3. But on the other hand if you put $a = 2$ and $b = -1$ then the left-hand side is 1 and the right-hand side is 3 so it's a clear $<$ in this case.

Secondly you might ask – why is (3.3.4) called the 'triangle inequality'? Well first of all, it's useful to think about (3.3.4) in terms of the real number line. The modulus of the real number x has a nice geometric meaning here – it is simply the *distance* of the number x from the 'origin' 0, or equivalently the *length* of the line segment joining 0 and x (see Figure 3.5). So (3.3.4) is telling us that the length of $a + b$ can never exceed the sum of the lengths of a and b.

Does this remind you of anything to do with triangles? If you know the result that in any triangle the length of the longest side can never be greater than the sum of the two other sides, then you're spot on. In fact that result can also be expressed in the form (3.3.4) when $|x|$ is re-interpreted as the length of a vector in two-dimensional space (see Figure 3.6).[5]

By the way, the interaction between the modulus and the $\times$ operation is much simpler and I'll leave it to you to show that for all real numbers a and b,

$$|ab| = |a||b|. \tag{3.3.5}$$

Note that as $\dfrac{a}{b} = a \times \dfrac{1}{b}$, we get $\left|\dfrac{a}{b}\right| = \dfrac{|a|}{|b|}$ as a free gift from (3.3.5).

Now we're ready to give the proof of Theorem 3.3.1. Because this result is so important we'll give not one but two proofs. Is this overdoing it? Well

---

[5] The inequality (3.3.4) also extends to higher and even infinite-dimensional spaces when $|x|$ is given a suitable meaning.

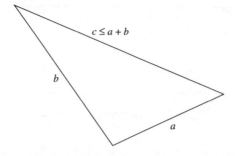

Figure 3.6. The triangle inequality in two dimensions.

mathematicians quite often have more than one proof of important results. A different proof can sometimes give new and important insights or might display greater elegance or beauty.[6]

Proof 1 of Theorem 3.3.1. This proof works by exhausting all four combinations of signs that a and b can have.

(i) $a \geq 0$, $b \geq 0$. In this case

$$|a + b| = a + b = |a| + |b|.$$

(ii) $a \leq 0$, $b < 0$. This works in more or less the same way as (i).

(iii) $a < 0$, $b \geq 0$. We then have $a = -|a|$. Now either $b > |a|$ or $b \leq |a|$.
If $b > |a|$, $|a + b| = b - |a| \leq b + |a| = |a| + |b|$.
If $b \leq |a|$, $|a + b| = |a| - b \leq |a| + |b|$.

(iv) $a > 0$, $b \leq 0$. This is proved the same way as in (iii).

The inequality (3.3.4) holds in all four cases. There are no other possibilities to consider and that completes the proof. □

Proof 2 of Theorem 3.3.1. Now since the square of a number is always nonnegative it follows that

$$|a + b|^2 = (a + b)^2$$
$$= a^2 + 2ab + b^2$$
$$= |a|^2 + 2ab + |b|^2.$$

---

[6] The great mathematician Carl Friedrich Gauss (1777–1855) gave eight distinct proofs in his lifetime of a result about numbers called the 'quadratic reciprocity theorem' and he published the first of these at the age of nineteen, see e.g. http://mathworld.wolfram.com/QuadraticReciprocityTheorem.html

Now using (3.3.3) and (3.3.5) we get

$$ab \leq |ab| = |a||b|,$$

and so we see that

$$|a + b|^2 \leq |a|^2 + 2|a||b| + |b|^2 = (|a| + |b|)^2.$$

Now use (L6) to get $|a + b| \leq |a| + |b|$ and our proof is complete. $\square$

Which proof do you prefer? Proof 2 is the one that usually appears in mathematical textbooks and it has the distinct advantage over Proof 1 of generalising directly to the higher-dimensional case, with the only necessary change being that $|x|$ is interpreted as the length of a vector.

We've now dealt with the interaction of the modulus with $\times, \div$ and $+$. The next question is what happens with $-$? I'm constantly surprised how many undergraduate students believe that the triangle inequality can be 'stretched' to give '$|a - b| \leq |a| - |b|$'. Of course this is WRONG! Just try $a = 2$ and $b = -3$ then the left-hand side is 5 while the right-hand side is -1. This at least suggests that maybe the inequality should be reversed so that we have $|a| - |b| \leq |a - b|$. That is correct, however it turns out that the order of $|a|$ and $|b|$ is unimportant here and that we also have $|b| - |a| \leq |a - b|$. Now since $||a| - |b||$ is either equal to $|a| - |b|$ or $|b| - |a|$ we might just as well prove the following – which is expressed as a corollary as it follows so easily from Theorem 3.3.1.

Corollary 3.3.1. For all real numbers a and b,

$$||a| - |b|| \leq |a - b|. \tag{3.3.6}$$

Proof. We'll use the mathematician's favourite trick (MFT) followed by the triangle inequality to get

$$|a| = |a - b + b|$$
$$\leq |a - b| + |b|.$$

Now use (L2) (with $c = -|b|$) to get

$$|a| - |b| \leq |a - b|.$$

Now repeat the argument that we've just used but with the roles of a and b interchanged. You then get

$$|b| - |a| \leq |b - a| = |a - b|,$$

and that's it. $\square$

3.4 Maxima and Minima

Here's another very short but useful little topic that finds its home in this chapter. Suppose that we have a list of N real numbers that I'll call $a_1, a_2, \ldots, a_N$. These are not in any particular order and the subscript $1, 2, \ldots, N$ is just used as a convenient device to distinguish the numbers from each other. Unless they are all equal to the same number, one of the numbers on the list will be the largest and one will be the smallest. The largest one is called the *maximum* and the smallest one is the *minimum*. We'll shorten these to max and min respectively so $\max(a_1, a_2, \ldots, a_N)$ picks out the largest number on the list and $\min(a_1, a_2, \ldots, a_N)$ identifies the smallest. To see how these are used in practice suppose that we have the following list: $-3, 1, 9, 4, -7$. Then

$$\max(-3, 1, 9, 4, -7) = 9 \quad \text{and} \quad \min(-3, 1, 9, 4, -7) = -7.$$

In general we have the following rather obvious inequalities: for all i running from 1 to N:

$$\min(a_1, a_2, \ldots, a_N) \le a_i \le \max(a_1, a_2, \ldots, a_N).$$

It's also worth pointing out a nice link between the modulus and the maximum:

$$|x| = \max(x, -x),$$

for any real number x.

3.5 The Theorem of the Means

In this section we'll give an example of a very interesting inequality which holds between two different types of average. Before we start to develop this it may be worth making a few remarks about *nth roots*. These are defined very similarly to square roots (see Section 2.2) but the number 'two' is replaced by an arbitrary natural number $n \ge 2$. So if x is a positive real number we define $\sqrt[n]{x}$ to be the unique positive real number y for which $y^n = x$. Indeed it can be shown that such a y always exists and if n is odd then y is the unique real number for which $y^n = x$. If n is even then $-y$ is also a solution to this equation. If x is a natural number, then just as was the case with square roots, $\sqrt[n]{x}$ is always an irrational number unless $x = a^n$ for some other natural number a, in which case $\sqrt[n]{x} = a$, e.g. $\sqrt[4]{81} = 3$ since $3^4 = 81$.

Now let's investigate averages. If you've attended just a very basic course in statistics you will have come upon the *arithmetic mean* which is often just called the *mean* or the *population mean*. To define it precisely let's suppose that we are

38

given n real numbers $a_1, a_2, \ldots, a_n$. Their arithmetic mean which I'll here denote by A_n is defined by

$$A_n = \frac{a_1 + a_2 + \cdots + a_n}{n}. \tag{3.5.7}$$

So for example if these were the heights of children in a school class, the arithmetic mean would be a measure of their average height and we might like to compare this to national data on heights for this age group to see if this group of kids was 'normal'.

The arithmetic mean is based on addition. By contrast, the *geometric mean* G_n is constructed using multiplication and we define this by

$$G_n = \sqrt[n]{a_1 a_2 \cdots a_n}. \tag{3.5.8}$$

Why do we call it geometric? Well if $n = 2$, $G_2 = \sqrt{a_1 a_2}$ is the length of the side of a square that has the same area as the rectangle with sides a_1 and a_2, if $n = 3$, $G_3 = \sqrt[3]{a_1 a_2 a_3}$ is the side of a cube that has the same volume as a box with sides a_1, a_2 and a_3, and so it continues to higher dimensions. It's interesting to compare G_n and A_n. First of all observe that if $a_1 = a_2 = \cdots = a_n = a$, then

$$A_n = \frac{na}{n} = a = \sqrt[n]{a^n} = G_n, \tag{3.5.9}$$

so the two means are always equal in this case. Now let's look at the case where $n = 5$ with $a_1 = 3, a_2 = 5, a_3 = 1, a_4 = 9$ and $a_5 = 11$. Here you can check that $A_5 = 5.8$ while $G_5 = 4.309$ (to 3 decimal places of accuracy). If you check with different sets of distinct numbers, you should find that the geometric mean is always smaller than the arithmetic mean. So that's our conjecture and we should aim to prove it.

Before we do this we will add one more weapon to our arsenal. We need to extend (L6) for squares to nth powers. This works fine and we have (the e here stands for 'extended'):

(L6e) Suppose that a and b are nonnegative real numbers. Then $a^n < b^n$ if and only if $a < b$.[7] The next result is the celebrated *Theorem of the Means*.

Theorem 3.5.1. The geometric mean of n real numbers can never exceed the arithmetic mean, i.e. given n real numbers $a_1, a_2, \ldots, a_n$

$$\sqrt[n]{a_1 a_2 \cdots a_n} \leq \frac{a_1 + a_2 + \cdots + a_n}{n}.$$

---

[7] To prove this you need the algebraic identity

$$b^n - a^n = (b - a)(b^{n-1} + ab^{n-2} + a^2 b^{n-3} + \cdots + a^{n-2}b + a^{n-1}).$$

Proof. To make our job simpler, we'll assume that all the a_is are unequal and that none of them are zero. In fact if any $a_i = 0$ then $0 = G_n < A_n$. I'll leave it to you to figure out how the proof I'll give below should be tweaked if two or more a_is are equal.

(*)Let $a_T = \max(a_1, a_2, \ldots, a_n)$ and $a_B = \min(a_1, a_2, \ldots, a_n)$. Now form a new set of numbers $a'_1, a'_2, \ldots, a'_n$. The only difference between these numbers and the original ones is that a_T has been removed and replaced by G_n and also a_B has been taken away and replaced by $\dfrac{a_T a_B}{G_n}$. Let G'_n and A'_n denote the geometric and arithmetic means (respectively) of this new set of numbers. Now a little bit of algebra should easily convince you that $G'_n = G_n$. It takes a bit more work to see how A'_n relates to A_n so let's do that now.

Since $a_B^n < a_1 a_2 \cdots a_n < a_T^n$ it follows from (L6e) that $a_B < G_n < a_T$. So $a_T - G_n > 0$ and $a_B - G < 0$. Multiplying these together we have

$$(a_T - G_n)(a_B - G_n) < 0$$

and expanding the bracket we get

$$a_T a_B - G_n(a_T + a_B) + G_n^2 < 0,$$

and if we rearrange this (using (L1) and (L2)) we obtain

$$a_T + a_B > \frac{a_T a_B}{G_n} + G_n.$$

Now notice that the left-hand side of this last expression is exactly the sum of the two terms we removed from the original list while the right-hand side is the sum of the two terms we replaced them with. When you substitute into the expressions for the two arithmetic means we see that this tells us that $A'_n < A_n$.(*)

Now I want you to think of the part of the proof that we've carried out so far and which is sandwiched between the two (*)s as a single procedure. The next thing to do is to carry out this procedure again so we let $a'_T = \max(a'_1, a'_2, \ldots, a'_n)$ and $a'_B = \min(a'_1, a'_2, \ldots, a'_n)$ and obtain a new set of numbers $a''_1, a''_2, \ldots, a''_n$ where a'_T has been removed and replaced by G_n and a'_B has been taken away and replaced by $\dfrac{a'_T a'_B}{G_n}$. If we repeat the reasoning that we carried out above we'll find that the arithmetic mean has changed to $A''_n < A_n$ while the new geometric mean is $G''_n = G_n$. Now after two applications of our procedure, two of the numbers on the original list have changed to G_n.

Maybe you can guess what happens next. Mathematicians call it *iteration* and use this to describe any procedure which is applied repeatedly and mechanically towards some goal. So after three applications of the procedure, the arithmetic mean will again be smaller, the geometric mean will stay the same and three numbers in the list will take the value G_n. Now we keep going until the nth stage. When we reach this stage, every number on the list is equal to G_n which is the arithmetic mean of that list by (3.5.9) but this is smaller than the arithmetic mean

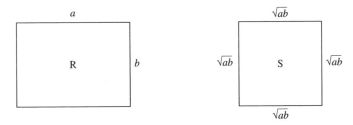

Figure 3.7. Square having the same area as a given rectangle.

at the $(n-1)$th stage which is smaller than that at the $(n-2)$th stage which is (working backwards through the iterations) certainly smaller than A_n. So we've shown that $G_n < A_n$, as was desired. □

The theorem we've just proved can be strengthened to show that $G_n = A_n$ if and only if all the numbers on the list are equal. We've proved part of this already (see (3.5.9)). You can prove the other part yourself by contemplating the way the proof of Theorem 3.5.1 works and thinking about what would happen if just two of the a_is were to be unequal.

The Theorem of the Means has an interesting geometric interpretation. Let $n = 2$ and consider the rectangle R with sides of length a_1 and a_2. The perimeter of R is $2(a_1 + a_2)$ and by Theorem 3.5.1 we have

$$2(a_1 + a_2) = 4\left(\frac{a_1 + a_2}{2}\right) \geq 4\sqrt{a_1 a_2}.$$

Now $4\sqrt{a_1 a_2}$ is precisely the perimeter of a square S having sides of length $\sqrt{a_1 a_2}$, as shown in Figure 3.7. Since both the square S and the rectangle R have the same area $a_1 a_2$ we see that the Theorem of the Means has told us that out of all possible rectangles that have the same area, the one with the smallest perimeter is the square. A similar geometric interpretation can be found in higher dimensions, see e.g. http://en.wikipedia.org/wiki/Inequality_of_arithmetic_and_geometric_means#Geometric_interpretation.

3.6 Getting Closer

One of our main reasons for learning about inequalities in this chapter (apart from their intrinsic interest) is so that we can use them in the next chapter in our study of limits. As a key step towards that goal we need to understand how mathematicians make sense of the concept of 'closeness'. To make this more precise, suppose we are given a fixed real number – let's call it l. We would like to say what it means for other numbers to be arbitrarily close to l without necessarily being equal to l.

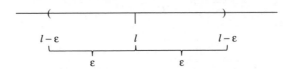

Figure 3.8. Interval of width 2ε centred on *l*.

To get a feel for this idea let's take $l = 1$. Suppose that I want to identify all the numbers that are within 0.01 of *l*. Then a moment's thought tells me that I need to be in the open interval (0.99, 1.01). Similarly if I wanted to be within 0.001 of 1 I would need the open interval (0.999, 1.001).

The numbers 0.01 and 0.001 are playing the role of 'degree of closeness' here. When we return to the case of a general real number *l* we will need a symbol to represent this degree of closeness and mathematicians have chosen the Greek letter, which is denoted by ϵ and pronounced epsilon, to fulfil this role. So if we want to get within ϵ of the real number *l* we must choose numbers that are in the open interval $(l - \epsilon, l + \epsilon)$ as shown in Figure 3.8, which is exactly what we did in the numerical examples where $l = 1$ and we took $\epsilon = 0.01$ and $\epsilon = 0.001$, respectively.

How can we express the fact that a real number x is close to *l* with ϵ measuring the degree of closeness? As we saw in Section 3.2 we must have

$$l - \epsilon < x < l + \epsilon. \tag{3.6.10}$$

Of course this is two inequalities rolled into one. The one on the left can be rearranged using (L2) to give $l - x < \epsilon$ and the one on the right similarly yields $x - l < \epsilon$, so we conclude that

$$|x - l| < \epsilon. \tag{3.6.11}$$

The fact that (3.6.10) and (3.6.11) are equivalent may just seem like pointless manipulation at the moment – but as the next chapter unfolds we will see how useful it is.

3.7 Exercises for Chapter 3

1. Show that if a, b and c are real numbers for which $a \leq b$ and $b \leq c$ then $a \leq c$. [Hint: Remember that $a \leq c$ means the same as $c - a \geq 0$. Now introduce b using the MFT.]

2. Prove that if a, b, c and d are real numbers satisfying $0 < a < b$ and $0 < c < d$ then $ac < bd$.

3. Find all values of x for which $\frac{x^2+1}{x+3} \leq 1$. [Hint: Treat the cases $x < -3$ and $x > -3$ separately.]

4. Find all values of x for which $\left|\frac{2x-1}{x+1}\right| < 1$. [Hint: Square both sides.]

5. Show that for all real numbers a, b, c, d with $|c| \neq |d|$ we have

$$\left|\frac{a+b}{c+d}\right| \leq \frac{|a|+|b|}{||c|-|d||}.$$

Do you think that either of the inequalities $\left|\frac{a+b}{c+d}\right| \leq \frac{|a|+|b|}{|c|+|d|}$ or $\left|\frac{a+b}{c+d}\right| \leq \frac{|a|+|b|}{|c|-|d|}$ is also true? Either present a proof or a counter-example in each case.

6. Let a and b be arbitrary real numbers. Use the fact that we always have $(a-b)^2 \geq 0$ to show that $ab \leq \frac{1}{2}(a^2+b^2)$. Hence deduce that $(a+b)^2 \leq 2(a^2+b^2)$. How do you think that this last inequality might generalise when the left-hand side is replaced by $(a_1 + a_2 + \cdots + a_n)^2$ for real numbers $a_1, a_2, \ldots, a_n$?

7. Use the binomial theorem (see Appendix 1) to prove *Bernoulli's inequality*:

$$(1+x)^r \geq 1 + rx,$$

where $x > 0$ and r is a natural number. For which natural numbers r is this inequality strict (i.e. $\geq$ can be replaced by $>$)? [If you know the technique of mathematical induction (see Appendix 3) you can try proving that the inequality holds for all $x > -1$.]

8. Consider the quadratic function

$$f(x) = ax^2 + bx + c,$$

where a, b and c are real numbers with $a > 0$. Show that

$$f(x) = a\left[\left(x + \frac{b}{2a}\right)^2 + \frac{4ac - b^2}{4a^2}\right],$$

and hence deduce that $f(x) \geq 0$ for all x if and only if $b^2 \leq 4ac$.

9. Recall the 'summation notation' or 'sigma notation'

$$\sum_{i=1}^{n} a_i = a_1 + a_2 + \cdots + a_n$$

(see Chapter 6 if you need to learn about this). A very famous and useful result is *Cauchy's inequality* for sums: If $a_1, a_2, \ldots, a_n$ and $b_1, b_2, \ldots, b_n$ are real numbers then

$$\left|\sum_{i=1}^{n} a_i b_i\right| \leq \left(\sum_{i=1}^{n} a_i^2\right)^{\frac{1}{2}} \left(\sum_{i=1}^{n} b_i^2\right)^{\frac{1}{2}}.$$

Verify that Cauchy's inequality is correct by applying the results of Exercise 8 to the quadratic function $f(x) = \sum_{i=1}^{n}(a_i x + b_i)^2$.

10. The next inequality may appear to be rather unexciting but we will use it in Chapter 5 to investigate square roots of prime numbers. Let $p > 0$ and let y be any other positive real number for which $y^2 \geq p$. Show that

$$p \leq \frac{1}{4}\left(y + \frac{p}{y}\right)^2 \leq y^2.$$

[Hint: The right-hand inequality is straightforward. For the left hand one, use the fact that if u and v are any two real numbers then

$$(u + v)^2 - 4uv = (u - v)^2 \geq 0.]$$

4

Where Do You Go To, My Lovely?

When the successive values attributed to a variable approach indefinitely a fixed value so as to end by differing from it by as little as one wishes, the last is called the limit of all the others.

A.-L. Cauchy quoted in *A History of Mathematics*, C.B. Boyer, U.C. Merzbach

4.1 Limits

In this section, we are going to meet the most important concept in this book. It may also (without fear of hyperbole) be one of the most essential ideas in the whole of mathematics. This beautiful and profound notion is the concept of *the limit* and its importance stems from the fact that it allows us to capture the infinite within the finite.

The limit pervades the subject of analysis but the simplest place to meet it is within the study of sequences. These are simply lists of numbers such as (S1) to (S4) which appear below:

(S1) $1, \frac{1}{2}, \frac{1}{3}, \frac{1}{4}, \frac{1}{5}, \frac{1}{6}, \frac{1}{7}, \cdots$

(S2) $1, 2, \frac{33}{13}, \frac{14}{5}, \frac{85}{29}, 3, \frac{161}{53}, \frac{52}{17}, \frac{261}{85}, \cdots$

(S3) $1, 1, 2, 3, 5, 8, 13, 21, 34, 55, 89, \ldots$

(S4) $1, 2, \frac{3}{2}, \frac{5}{3}, \frac{8}{5}, \frac{13}{8}, \frac{21}{13}, \frac{34}{21}, \frac{55}{34}, \frac{89}{55}, \cdots$

Each of these shows the beginning of a list that extends indefinitely. Of course there are many (an infinite number of!) ways to carry on each list in each case, but in mathematics we usually expect that there is a pattern expressed by a formula that enables us to calculate any term on the list (at least in principle). The pattern for (S1) is pretty easy to identify. In this case the first term on the list is $\frac{1}{1}$, the second term is $\frac{1}{2}$, the third term is $\frac{1}{3}$ and so on. Hence the nth term is $\frac{1}{n}$ and we can assert immediately (without having to count along to that point) that the 1024th term is $\frac{1}{1024}$.

The pattern in (S2) is more difficult to discern and that's probably because it's one that I manufactured myself by starting with the formula for the nth term $\frac{2n+3n^2}{n^2+4}$. So for example, to get the 6th number on the list I put $n = 6$ into this formula to get $\frac{12+108}{36+4} = \frac{120}{40} = 3$. The third and fourth lists are more famous and interesting examples. (S3) Is the Sequence of *Fibonacci numbers*. It was studied by Leonardo of Pisa (c.1180–1250) (who was also known as Fibonacci) and appears in his book *Liber Abaci* which was published in 1202.[1] It describes an idealised model of how a population of rabbits might grow. In this case the nth term in the sequence is the number of pairs of rabbits at the end of the nth month. So if we start with one pair of rabbits (one male and one female) at the beginning then there is still one pair at the end of both the first and second months, but by the end of the third month these rabbits have bred and the population doubles. There is a very simple formula that can be used from this point on to calculate f_n, the number of pairs of rabbits at the end of month n, from the numbers at the end of the previous two months. It is

$$f_n = f_{n-1} + f_{n-2}, \tag{4.1.1}$$

so for example, I finished the list in (S3) with $f_{10} = 55$ and $f_{11} = 89$ from which I can then calculate $f_{12} = 55 + 89 = 144$ and $f_{13} = 89 + 144 = 233$. Incidentally any sequence like (4.1.1) in which later values are calculated in terms of earlier ones is called a *recursion*.

The sequence (S4) is closely related to (S3). We won't say much about it now but we will come back to it later as it is associated with a very famous number – the golden section (see page 22) and we'll get to that number by taking limits! For now it will suffice to just spot the general formula, so if r_n in the nth term is the sequence then you can check that it is the ratio of successive Fibonacci numbers, i.e.

$$r_n = \frac{f_{n+1}}{f_n}. \tag{4.1.2}$$

We need to be able to develop a precise mathematical procedure for working with sequences. We will formally define a *sequence* to be a list of real numbers that is indexed by the natural numbers.[2] The generic sequence will be denoted by (a_n) and the number which appears as the nth term is a_n. Please do not get confused between (a_n) and a_n. (a_n) is shorthand for the list of numbers $a_1, a_2, a_3, \ldots$ which

---

[1] See e.g. http://en.wikipedia.org/wiki/Fibonacci#Fibonacci_sequence

[2] This is a working definition which is fine for this book – but a more precise definition is that a sequence is a mapping (or function) from the (set of) natural numbers to the (set of) real numbers.

Figure 4.1. The harmonic sequence.

goes on forever. I stress that it is a list and not a number. On the other hand a_n really is a number, e.g. in (S2), $a_4 = \frac{14}{5} = 2.8$.

The key question we are now going to turn our attention to is – what happens to a sequence (a_n) as n gets very large? That's a vague question and we'll need to explore a little further before we try to answer it. We begin by looking more closely at the sequence (S1). It is called the *harmonic sequence* (Figure 4.1). This terminology comes from the relationship between numbers and music that goes back at least as far as the Pythagoreans.[3] They observed that if you pluck e.g. a guitar string at a $\frac{1}{2}, \frac{1}{3}, \frac{1}{4}$ etc. of its length then the pitch increases and you create a succession of notes with higher frequencies. The terms in the sequence look as though they are getting closer and closer to zero as n gets larger and larger but observe that it can never reach zero. Indeed if there was a natural number N which had the property that $\frac{1}{N} = 0$ we could multiply both sides by N to see that $1 = 0$, which is a contradiction.

When we look at (S2), we see that both the numerator and the denominator of $a_n = \frac{2n+3n^2}{n^2+4}$ get larger and larger as n grows so it appears that the sequence is being attracted towards the 'number' $\frac{\infty}{\infty}$ which we know to be undefined! On the other hand, we can calculate $a_{12} = \frac{456}{148} = 3.08108\ldots$, $a_{100} = \frac{30200}{10004} = 3.01879\ldots$ which provides at least some evidence that we should interpret $\frac{\infty}{\infty}$ as 3 in this case. In (S3) the story seems much simpler, as n gets larger then so does f_n, but (S4) seems to be behaving more like (S1) and (S2) and you should check that the terms appear to be closing in on a number that is in the region of 1.618.

We will now try to pin down more clearly the behaviour that we've identified in (S1), (S2) and (S4). We'll introduce two new terms. The first of these is the idea of *convergence* and at this stage we will say that the sequence (a_n) converges if when n gets very large, the term a_n gets arbitrarily close to a real number l. In this case we will call l the *limit* of the sequence (a_n). We have already seen in (S1) that the limit (which is 0 in that case) may never be reached, but what does happen there is that as n gets larger and larger we get *arbitrarily closer* to 0. We have not yet given adequate mathematical definitions of the concepts of convergence and limit. To achieve this we need to capture the notion of arbitrary closeness more precisely. We came some way towards doing that in Section 3.6, so if we fix a degree of closeness ϵ then it seems that $|a_n - l| < \epsilon$ is what we are looking for. But do we want this to hold for all values of n? If we do this in (S1) and take $n = 1$

---

[3] This term usually refers to followers of the mathematician/philosopher Pythagoras who were active in the period from about 585 to 400 BCE.

and we believe that $l = 0$ we get $|a_1 - l| = 1 - 0 = 1$, which isn't very small. So what is missing from our definition? Well if you think about it I hope you'll agree that we can't have 'arbitrary closeness' unless we have journeyed far enough along the sequence. How do we express 'far enough along' mathematically? How about this? Suppose we fix a natural number n_0 as a starting point and argue that for all $n > n_0$ we must have arbitrary closeness. This will nearly work but what is still missing is a connection between the two numbers: ϵ that measures closeness and n_0 which identifies the point at which closeness kicks in. But maybe it's time now to stop this discussion and give the definition which we've been building up to. This is the key definition in this book, so let's concentrate very carefully on what it says:

> The sequence (a_n) *converges* to the real number l if given any $\epsilon > 0$ there exists a natural number n_0 so that whenever $n > n_0$ we must have $|a_n - l| < \epsilon$.

If such a number l exists we call it the *limit* of the sequence (a_n) and we write

$$l = \lim_{n \to \infty} a_n. \tag{4.1.3}$$

An alternative (and completely equivalent) notation to (4.1.3) is

$$a_n \to l \quad \text{as} \quad n \to \infty.$$

This is a subtle and powerful definition. It takes some time to internalise so don't worry if you feel like you don't understand it yet. I can assure you that if this is the case then you are in very good company. I also want to stress that the notation $n \to \infty$ is a symbolic one that is supposed to be suggestive of approaching infinity but not of reaching it (whatever that means) so that the problems that we encountered in Section 2.4 are irrelevant here. The definition of convergence links the two numbers ϵ and n_0 in the following way. As ϵ measures degree of closeness we can take ϵ to be as small as we like, but the smaller we take ϵ, the larger we must take n_0 as we then have to go further along the sequence to find those numbers n for which $|a_n - l| < \epsilon$.

For example consider (S1) again and let $\epsilon = 0.12$. To get $|a_n - l| = \left| \frac{1}{n} - 0 \right| < \epsilon$ in this case, we require $\frac{1}{n} < 0.12$, i.e. $n > \frac{1}{0.12} = 8.333\ldots$ So here we can take $n_0 = 8$ and the definition is satisfied – but only for this particular choice of ϵ. If we instead take $\epsilon = 0.012$, you can check that we then need n_0 to be at least 83 and $n_0 = 833$ or more is required if $\epsilon = 0.0012$. Do these numerical calculations enable us to prove convergence? No they don't, as the definition requires us to consider *all* $\epsilon > 0$ and this is an infinite number of cases to consider. Playing with specific numbers helps us to get a feel for how the definition works, but to legitimately prove convergence we need to be cleverer!

Example 4.1: *To show* $\lim_{n \to \infty} \frac{1}{n} = 0$.

We begin with the first line of the definition and choose an arbitrary $\epsilon > 0$. Now consider $\frac{1}{\epsilon}$. This is a real number (which is large when ϵ is small) and so it has a decimal expansion $\frac{1}{\epsilon} = n_0.n_1 n_2 n_3 \cdots$. So $\frac{1}{\epsilon} \geq n_0$ and by (L5) we have $\frac{1}{n_0} \leq \epsilon$. So given any $\epsilon > 0$, if $n > n_0$ we have

$$\left| \frac{1}{n} - 0 \right| = \frac{1}{n} < \frac{1}{n_0} \leq \epsilon,$$

i.e. $\frac{1}{n} < \epsilon$ so this choice of n_0 satisfies the definition and we may 'legitimately' write $\lim_{n \to \infty} \frac{1}{n} = 0$.

I've put the word legitimately in quotes as although this 'proof' will do for now, it isn't really good enough. That's because we've had to use a working definition of a real number as one that has a decimal expansion in order to extract our n_0 from our ϵ. To give a fully rigorous proof we'll need a more sophisticated understanding of real numbers and this must be postponed for now. The good news is that none of the general proofs in this chapter will require this deeper insight into the real line and they are completely rigorous as they stand. In Chapter 11 we will prove the *Archimedean property* of the real numbers and that gives the firm foundation that we need for Example 4.1.

The first general theorem that we'll look at concerns *uniqueness*. Mathematicians worry quite a lot about this concept and if you think about it, they are right to. If a sequence were to converge to more than one limit – then which one is the right one? Fortunately, as the theorem below shows – that can never be an issue.

Theorem 4.1.1. If a sequence converges to a limit, then that limit is unique. More precisely if (a_n) is a sequence such that $\lim_{n \to \infty} a_n = l$ and $\lim_{n \to \infty} a_n = l'$ then $l = l'$.

Proof. We'll use a proof by contradiction, Suppose that l and l' are as in the statement of the theorem with $l \neq l'$. By definition of convergence, given any $\epsilon > 0$ there exists a natural number n_0 such that if $n > n_0$ then $|a_n - l| < \frac{\epsilon}{2}$, (why $\frac{\epsilon}{2}$ and not ϵ? We'll return to this point in the discussion after the proof) and there also exists a natural number m_0 such that if $n > m_0$ then $|a_n - l'| < \frac{\epsilon}{2}$. Now we'll use MFT (the mathematician's favourite trick) and the triangle inequality. Ensure that $n > \max(m_0, n_0)$ so that we have closeness to both limits. Then

$$|l - l'| = |l - a_n + a_n - l'|$$
$$\leq |l - a_n| + |a_n - l'|$$
$$< \frac{\epsilon}{2} + \frac{\epsilon}{2} = \epsilon.$$

So we've shown that for any $\epsilon > 0$, $|l - l'| < \epsilon$ so that $|l - l'|$ is smaller than any positive number. By properties of the modulus we also know that $|l - l'| \geq 0$ and so the only possibility is that $|l - l'| = 0$, i.e. $l = l'$ and this is our required contradiction. $\qquad \square$

It's a good idea to read through the proof of Theorem 4.1.1 a few times to make sure you really understand it. This will serve you in good stead for later as it contains many features that are typical in analysis proofs. You should understand why m_0 and n_0 cannot be chosen to be the same (if l and l' really were different, you may need to go further along the sequence in one case than the other in order to reach the desired degree of closeness). Finally why did I use $\frac{\epsilon}{2}$ instead of ϵ? Well first of all, there is a sense in which it doesn't really matter. Remember that the definition of a limit l of a sequence (a_n) tells us that given *any* $\epsilon > 0$ there exists n_0 such that if $n > n_0$ then $|a_n - l| < \epsilon$. Now we can certainly replace ϵ here with $\frac{\epsilon}{2}$ (or indeed $K\epsilon$ for any fixed real number K) since being 'given any $\epsilon > 0$' is the same as being 'given any $\frac{\epsilon}{2} > 0$' (think about it!) Of course $\frac{\epsilon}{2}$ is smaller than ϵ so imposing a smaller degree of closeness requires us to choose larger m_0 and n_0 – but so what? The reason for using $\frac{\epsilon}{2}$ rather than ϵ is one of mathematical style and elegance. If we'd used ϵ then we'd have finished the proof with $|l - l'| < 2\epsilon$ and this looks ugly to the trained mathematician. Doing Exercise 4.9 at the end of this chapter will help you to clarify this issue.

Let's move on from these finicky points to consider another example. In the following we'll fix $-1 < r \leq 1$.

Example 4.2: *To show* $\lim_{n \to \infty} r^n = 0$ (if $-1 < r < 1$).

Before we go further, let's look at a concrete example and take $r = \frac{1}{2}$. As n gets larger so does 2^n and so $\frac{1}{2^n}$ gets smaller and smaller so the result that is claimed looks feasible.

To solve the problem we need to apply a case-by-case method:

Case 1: $0 < r < 1$. To prove the result in this case, we note that as $r < 1$, we have $\frac{1}{r} > 1$ by (L5). It turns out to be a good idea to write $\frac{1}{r} = 1 + h$ where $h > 0$. Then using the binomial theorem (see Appendix 1 if you need some background on this) we get

$$\frac{1}{r^n} = \left(\frac{1}{r}\right)^n = (1 + h)^n = 1 + nh + \frac{1}{2}n(n - 1)h^2 + \cdots + h^n,$$

so that in particular – since all terms on the right-hand side are positive, we have[4]

$$\frac{1}{r^n} > nh$$

---

[4] If you don't want to use the binomial theorem it's enough to notice that

$$(1 + h)^n = \underbrace{(1 + h)(1 + h)\cdots(1 + h)}_{n \text{ times}} > nh$$

as when you multiply out the n brackets you always get a term where the 1 in $(n - 1)$ of the products of $(1 + h)$ meets the h in the final $(1 + h)$ and there are n ways in which this happens. Note that we have essentially proved Bernoulli's inequality from Exercise 3.7.

and so by (L5) again

$$r^n < \frac{1}{nh} = \frac{1}{h} \cdot \frac{1}{n}.$$

In Example 4.1, we showed that $\lim_{n\to\infty} \frac{1}{n} = 0$, and so given any $\epsilon > 0$, there exists n_0 such that if $n > n_0$ then $\frac{1}{n} < \epsilon h$.[5] So if $n > n_0$ we see that

$$r^n < \frac{1}{h} \cdot \epsilon h = \epsilon,$$

and the proof is complete.

Case 2: $-1 < r < 0$. This is easy. We just repeat the above reasoning with r replaced throughout by $|r|$ as $0 < |r| < 1$.

Case 3: $r = 0$. If $r = 0$ then $r^n = 0$ and it's very easy to prove that $\lim_{n\to\infty} 0 = 0$ – indeed something would be very wrong with the definition of convergence if this wasn't so. This completes the proof.

We can generalise the argument of case 3 to show that the constant sequence which is such that $a_n = c$ for all n, where c is a fixed real number, converges to c. So we see that $\lim_{n\to\infty} c = c$. In particular if we apply this to the case $c = 1$ we can extend the convergence of the sequence (r^n) to allow $r = 1$, but in this case the limit is 1 and not 0. By symmetry, we might expect to be able to include $r = -1$ as well, but the argument breaks down here as $(-1)^n = 1$ when n is even and is -1 when n is odd. So the sequence is $-1, 1, -1, 1, -1, 1, \ldots$ and there can be no possibility of convergence to a limit.

We haven't looked systematically yet at sequences that fail to converge and this is an opportunity too good to miss. Here's a formal definition – a sequence is said to *diverge* if it doesn't converge. This is not so much a definition, as a piece of terminology. The name 'divergence' has stuck in the literature but seems like something of a misnomer as intuitively we'd expect divergence to mean 'inexorably moving away' or 'getting larger in size'. In fact divergence can be subdivided into more than one type of behaviour and in two cases, there is some reconciliation with intuition. Before we give the definitions, let's consider two examples. The sequence $(n) = 1, 2, 3, 4, 5, \ldots$ is just the natural numbers. They grow inexorably larger and we know there is no end-point. To capture this type of behaviour we say that a sequence (a_n) *diverges to* $+\infty$ if, given any real number $K > 0$, there exists a natural number n_0 such that $a_n > K$ for all $n \geq n_0$. This is rather like the definition of convergence except that then we were dealing with degree of closeness and we required a number ϵ that could be made arbitrarily small. In place of ϵ we now have K that can be made arbitrarily large, but no

---

[5] The fixed number h will play exactly the same role here that $\frac{1}{2}$ did in the proof of Theorem 4.1.1.

matter how large it is taken, if we go far enough along the sequence we can find a point after which all the a_ns exceed K. Notice that the use of ∞ in the definition is just as part of a name. There are no infinite numbers involved in the definition itself. But what about the sequence $(-n)$ of negative integers? It clearly diverges and has a similar behaviour to n but it is going in the wrong direction for the definition we've just given. In order to accomodate this sort of behaviour, we say that a sequence (b_n) *diverges to* $-\infty$ if given any real number $L > 0$ there exists a natural number m_0 such that $b_n < -L$ for all $n \geq m_0$. Finally we say that a sequence is *properly divergent* if it diverges to either $= \infty$ or $-\infty$.

What about the sequence $(-1)^n$? It diverges but certainly not to $\pm\infty$. A sequence that displays similar behaviour, i.e. endlessly cycles between two or more numbers is said to *oscillate*. In the case of $(-1)^n$, we see that it *oscillates finitely* as there are only a finite number of possibilities (two in this case). We'll see an example below of a sequence that *oscillates infinitely*.[6]

Example 4.3: The divergence of the sequence (r^n) if $r \leq -1$ or $r > 1$.

We will show that (r^n) diverges to $+\infty$ if $r > 1$. We do this by a similar argument to that used in Case 1 of Example 4.2. We write $r = 1 + h$ where $h > 0$. Using the binomial theorem as before (or arguing directly as in footnote 4 within Example 4.2) we conclude that $r^n = (1 + h)^n > nh$. Now choose $K > 0$ to be as large as you like and write the real number $\frac{K}{h} = m_0.m_1 m_2 m_3 \cdots$ We take $n_0 = m_0 + 1$, then for $n > n_0$, we have

$$r^n > n_0 h = (m_0 + 1)h > \frac{K}{h}.h = K,$$

and we are done.

If $r = -1$, we have already seen that the sequence oscillates finitely. Finally if $r < -1$, the sequence oscillates infinitely. We won't prove this but if you take e.g. $r = 2$, you see that $(-2)^n = -2, 4, -8, 16, -32, \ldots$ and the behaviour is quite clearly different from that of $(-1)^n$.

Even in as simple a sequence as (r^n) we see that we encounter all but one of the different notions of divergence as well as convergence, when we vary the value of r. It's perhaps worth listing the results of Examples 4.2 and 4.3 so that we can examine them together.

4.1.1 The story of the sequence (r^n).

- If $r < -1$, the sequence oscillates infinitely.
- If $r = -1$, the sequence oscillates finitely.
- If $-1 < r < 1$, the sequence converges to 0.

---

[6] Precise definitions of these concepts are given in the next section.

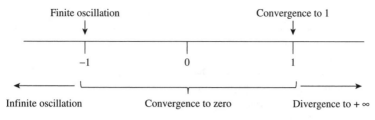

Figure 4.2. The story of (r^n).

- If $r = 1$, the sequence converges to 1.
- If $r > 1$, the sequence diverges to $+\infty$.

We can think of the behaviour of this *dynamical system* from a different point of view. We are varying the *parameter* r over the whole real line and it divides that line into different regions corresponding to how the sequence (r^n) behaves *asymptotically* (i.e. for large n). The isolated points $r = -1$ and $r = 1$ are quite interesting here (see Figure 4.2) as in each case they are boundaries between two very different regions of behaviour, e.g. $r = 1$ is the boundary between the region $0 < r < 1$ (convergence to zero) and $r > 1$ (divergence to $+\infty$). The behaviour at the boundary is completely different (in each case) from that in either of the two regions that it separates and this is not untypical of far more complicated systems.

At this stage it might have occurred to you that we can use limits of sequences to make rigorous sense of what we mean by irrational numbers, by building these systematically as limits of sequences of rational numbers. For example $\sqrt{2}$ should be the limit of a sequence that starts off as $1, 1.4, 1.41, 1.414, 1.4142, 1.41421, 1.414213, 1.4142135, 1.41421356, \ldots$[7] This is indeed the case and the real number line can be given a very concise and mathematically satisfying meaning as the *completion* of the rational numbers through taking limits of sequences. The details are sophisticated. We'll leave them for now and come back to this important idea in Chapter 11.

We'll finish this section with two useful results. We'll first show that the limit of a sequence of positive numbers can never be negative.

Theorem 4.1.2. Let (a_n) be a sequence with each $a_n \geq 0$ and suppose that the sequence converges to l. Then $l \geq 0$.

Proof. Assume that $l < 0$. As (a_n) converges to l we know that given any $\epsilon > 0$ there exists a natural number n_0 such that if $n > n_0$ then $|a_n - l| < \epsilon$, i.e. $l - \epsilon < a_n < l + \epsilon$ (recall (3.6.10) and (3.6.11)). Now ϵ can be any positive

---

[7] You can extract the first million terms in this sequence by visiting http://antwrp.gsfc.nasa. gov/htmltest/gifcity/sqrt2.1mil

number we like so let's take $\epsilon = \frac{|l|}{2} = -\frac{l}{2}$. Then for $n > n_0$ we have $a_n < l - \frac{l}{2} = \frac{l}{2}$ which is negative. So we've proved that a_n is negative for all $n > n_0$ which is a contradiction and so we conclude that $l \geq 0$, as required. □

Like so many results in analysis, Theorem 4.1.2 is delicate. If you change $a_n \geq 0$ to $a_n > 0$ you might at first expect that $l \geq 0$ should change to $l > 0$. This isn't true – for a counter-example (i.e. an example that by its very existence disproves the claim) consider $a_n = \frac{1}{n}$.

The next result is a useful one about comparing two divergent sequences. We'll omit the proof as it follows almost immediately from the definition and you should check this.

Theorem 4.1.3. Suppose that (a_n) and (b_n) are two sequences with $a_n \leq b_n$ for all n.

1. If (a_n) diverges to $+\infty$ then so does (b_n).

2. If (b_n) diverges to $-\infty$ then so does (a_n).

By the way, now we know what limits really mean we can return to a theme we discussed at the end of Section 1.2. We were trying to understand the behaviour of $\pi(n)$ – the number of prime numbers less-than or equal to n and I told you about the celebrated *prime number theorem* – but only in a very vague way. Now I can state this theorem precisely. It says very succinctly that

$$\lim_{n \to \infty} \frac{\pi(n) \log_e(n)}{n} = 1,$$

so that given any $\epsilon > 0$ there exists a natural number N such that for all $n > N$, $\left| \frac{\pi(n) \log_e(n)}{n} - 1 \right| < \epsilon$. In Section 1.2, I was misleading you as I asked you to look at the behaviour of $\pi(n) - \frac{n}{\log_e(n)}$ and you can now see why this is incorrect. $\frac{n}{\log_e(n)}$ is a good approximation to $\pi(n)$ in the sense that $\pi(n) \div \frac{n}{\log_e(n)}$ gets closer and closer to 1 as n gets larger and larger. You should test this claim out with some numbers. Regrettably, we can't prove the theorem here as it uses techniques that go beyond the scope of this book.

4.2 Bounded Sequences

Before we prove any more theorems we need a new concept. We say that a sequence (a_n) is *bounded* if there exists a real number $K > 0$ such that

$$|a_n| \leq K \text{ for all natural numbers } n,$$

i.e. we always have $-K \leq a_n \leq K$ so that all the terms of the sequence are trapped between $-K$ and K. Notice that there is nothing special (at this point) about the number K in this definition. Once we've found a K that works then any larger number will also do the trick. (S1) is an example of a bounded sequence. Here we can take any $K \geq 1$. (S2) is certainly not bounded. We'll come back to (S2) and (S4) later. We've already argued that (S1) is convergent and we've just pointed out that it is bounded. But there are many examples of bounded sequences that are not convergent, e.g. consider the sequence whose nth term is $(-1)^n$. We know that it diverges but it is clearly bounded (again just take any $K \geq 1$). Any sequence that fails to be bounded is said to be *unbounded*. There is an important relationship between convergent and bounded sequences which is given in the following theorem.

Theorem 4.2.1. If a sequence (a_n) is convergent then it is also bounded.

Proof. We need to find $K > 0$ such that $|a_n| \leq K$ for all n. But we know (a_n) converges to some real number l so given any $\epsilon > 0$ there exists a natural number n_0 so that if $n > n_0$ then $|a_n - l| < \epsilon$. Now by MFT and the triangle inequality, if $n > n_0$

$$|a_n| = |(a_n - l) + l|$$
$$\leq |a_n - l| + |l|$$
$$< \epsilon + |l|.$$

So we can take $K = \epsilon + l$ provided $n > n_0$. We need a K that works for all n so now suppose that $n \leq n_0$. We then have

$$|a_n| \leq \max(|a_1|, |a_2|, \ldots, |a_{n_0}|).$$

Now if we combine together the two pieces of our argument we see that a K which works for all n is

$$K = \max(|a_1|, |a_2|, \ldots, |a_{n_0}|, \epsilon + |l|),$$

and that completes the proof. □

Theorem 4.2.1 tells us that the convergent sequences 'sit inside' the bounded sequences.[8] If you reverse the logic in the statement of the theorem then you see that an unbounded sequence must diverge. This is often the quickest way to see that a sequence is divergent and (S3) is a case in point here.

---

[8] If you know about sets then the set of all convergent sequences is a subset of the set of all bounded sequences. See Appendix 2.

We can use the concept of a bounded sequence to make the classification of divergent sequences much more precise. We say that a sequence

- *oscillates* if it is neither convergent nor properly divergent.
- *oscillates finitely* if it oscillates and is bounded.
- *oscillates infinitely* if it oscillates and is not bounded.

Now that we have this definition it should be an easy exercise for you to prove that r^n oscillates infinitely when $r < -1$. Note that a sequence can oscillate finitely but still take on infinitely many different values, e.g. consider the sequence whose nth term is $(-1)^n \frac{1}{n}$. The word 'finitely' here is really indicating the boundedness of the sequence and the fact that there is some finite interval from which it can never escape (which is $[-1, \frac{1}{2}]$ for the last example).

We'll finish this section with a useful little result. First a definition. If a sequence (a_n) converges to zero then it is called a *null sequence*.

Theorem 4.2.2. If (a_n) is a null sequence and (b_n) is bounded then the sequence $(a_n b_n)$ is also a null sequence.

Proof. We need to show that $(a_n b_n)$ converges to zero. Since (b_n) is bounded there exists $K > 0$ such that $|b_n| < K$ for all n. On the other hand since (a_n) converges to zero, given any $\epsilon > 0$ there exists n_0 such that for all $n \geq n_0$, $|a_n| = |a_n - 0| < \frac{\epsilon}{K}$. But then for such n

$$|a_n b_n| = |a_n|.|b_n| < \frac{\epsilon}{K}.K = \epsilon,$$

and that is what is needed to prove the result. □

As an example, we find that $(-1)^n \frac{1}{n}$ is a null sequence since the sequence $(-1)^n$ is bounded. If you know about trigonometry, then Theorem 4.2.2 can also be used to show that sequences like $a_n = \frac{\cos(n)}{n^2}$ are null sequences.

Beware that if (a_n) converges to a non-zero limit, for (b_n) to be bounded is not enough to guarantee convergence of $(a_n b_n)$, e.g. take $a_n = 1 - \frac{1}{n}$ which converges to 1 and $b_n = (-1)^n$. In this case $a_n b_n$ oscillates finitely.

4.3 The Algebra of Limits

Let's return to the sequence (S2) whose nth term is $\frac{2n+3n^2}{n^2+4}$. We presented some evidence earlier that the limit might be 3. How would we prove this properly? First of all we'll do a little bit of algebra that has nothing to do with analysis. We'll divide every term in the numerator and denominator of the fraction by the highest power of n that occurs. This is n^2 and we get:

$$a_n = \frac{2n + 3n^2}{n^2 + 4} = \frac{\frac{2}{n} + 3}{1 + \frac{4}{n^2}}.$$

Now we know that $\lim_{n \to \infty} \frac{1}{n} = 0$, $\lim_{n \to \infty} 3 = 3$, $\lim_{n \to \infty} 1 = 1$ and $\lim_{n \to \infty} 2 = 2$. We could indeed argue that $\lim_{n \to \infty} a_n = 3$ if we could justify writing

$$\lim_{n \to \infty} a_n = \frac{2 \lim_{n \to \infty} \frac{1}{n} + 3}{1 + 4 \lim_{n \to \infty} \frac{1}{n} \lim_{n \to \infty} \frac{1}{n}}.$$

It turns out that this sort of reasoning is indeed justified and the general result that we need is given in the next theorem – which is often known as *the algebra of limits*.

Theorem 4.3.1. Suppose that (a_n) and (b_n) are convergent sequences with $\lim_{n \to \infty} a_n = l$ and $\lim_{n \to \infty} b_n = m$ then

1. The sequence whose nth term is $a_n + b_n$ converges to $l + m$.

2. The sequence whose nth term is $a_n b_n$ converges to lm.

3. If c is any real number then the sequence whose nth term is ca_n converges to cl.

4. If $b_n \neq 0$ for all n and also $m \neq 0$ then the sequence whose nth term is $\frac{a_n}{b_n}$ converges to $\frac{l}{m}$.

Proof.

1. This is fairly similar to that of Theorem 4.1.1 and it goes like this. Given any $\epsilon > 0$, we know that there exists a natural number n_0 such that if $n > n_0$ then $|a_n - l| < \frac{\epsilon}{2}$ and there also exists a natural number p_0 such that if $n > p_0$ then $|b_n - m| < \frac{\epsilon}{2}$. Now choose $n > \max(n_0, p_0)$ and apply the triangle inequality to see that

$$|(a_n + b_n) - (l + m)| = |(a_n - l) + (b_n - m)|$$
$$\leq |a_n - l| + |b_n - m|$$
$$< \frac{\epsilon}{2} + \frac{\epsilon}{2} = \epsilon.$$

2. Here we'll use the MFT and the triangle inequality in what I hope is now a familiar way – but we'll also need to appeal to Theorem 4.2.1. First we have

$$|a_n b_n - lm| = |a_n b_n - lb_n + lb_n - lm|$$
$$\leq |b_n(a_n - l)| + |l(b_n - m)|$$
$$= |b_n||a_n - l| + |l||b_n - m|, \dots (*)$$

where we have used (3.3.5) to get the last line. At this stage we'll assume that $l \neq 0$ and worry about what happens when $l = 0$ later on. Now the sequence (b_n) is convergent and so by Theorem 4.2.1 is bounded. Hence there exists a real number $K > 0$ such that $|b_n| < K$ for all n. So we can go back to (*) and write

$$|a_n b_n - lm| \leq K|a_n - l| + |l||b_n - m|. \ldots (**)$$

Now choose $\epsilon > 0$, then there exists n_0 such that if $n > n_0$, $|a_n - l| < \frac{\epsilon}{2K}$ and there exists p_0 such that $|b_n - m| < \frac{\epsilon}{2|l|}$ whenever $n > p_0$. From (**) we then get for $n > \max(n_0, p_0)$,

$$|a_n b_n - lm| < K . \frac{\epsilon}{2K} + |l| . \frac{\epsilon}{2|l|}$$

$$= \frac{\epsilon}{2} + \frac{\epsilon}{2} = \epsilon.$$

That proves the theorem in the case where $l \neq 0$. If $l = 0$ then just go back to (*) and use the fact that given $\epsilon > 0$ there exists r_0 such that if $n > r_0$, $|a_n| < \frac{\epsilon}{K}$.

3. This follows from the result (2) that we've just proved by taking (b_n) there to be the constant sequence whose nth term is the real number c.

4. This proof is a bit finicky so we'll do it in stages. First we'll do the hard part and show that $\lim_{n \to \infty} \frac{1}{b_n} = \frac{1}{m}$. Now this will mean that we'll have to look at terms like

$$\left| \frac{1}{b_n} - \frac{1}{m} \right| = \left| \frac{m - b_n}{mb_n} \right|$$

$$= \frac{1}{|b_n|} . \frac{1}{|m|} . |b_n - m| \ldots (\dagger)$$

Now for large enough n we can make $|b_n - m|$ as small as we like and $\frac{1}{|m|}$ is constant and so presents no problems. The problem term in ($\dagger$) is $\frac{1}{|b_n|}$ so let's focus on that. To deal with this we'll need to be clever and we'll choose $\epsilon < \frac{|m|}{2}$. As we continue the argument, you'll see why this is a good idea. Given such an ϵ we can as usual find n_0 such that if $n > n_0$ then

$$|b_n - m| < \epsilon < \frac{|m|}{2}.$$

So by Corollary 3.3.1, we have

$$|m| - |b_n| < \frac{|m|}{2},$$

and so by (L2), $|b_n| > \frac{|m|}{2}$. Now we can use (L5) to deduce that

$$\frac{1}{|b_n|} < \frac{2}{|m|}.$$

We can then use the same argument as in the proof of Theorem 4.2.1 to see that the sequence whose nth term is $\frac{1}{|b_n|}$ is bounded with $K = \max\left(\frac{1}{|b_1|}, \frac{1}{|b_2|}, \ldots, \frac{1}{|b_{n_0}|}, \frac{2}{|m|}\right)$.

Now let's return to (†). With ϵ as chosen we see that for $n > n_0$ we have

$$\left| \frac{1}{b_n} - \frac{1}{m} \right| < \frac{K}{|m|}\epsilon,$$

and that will suffice to establish that $\lim_{n\to\infty} \frac{1}{b_n} = \frac{1}{m}$. Finally, to show the general result claimed in the theorem, we just write $\frac{a_n}{b_n} = a_n \cdot \frac{1}{b_n}$ and use the result of (2) that was proved above. $\qquad\square$

In the proof we've just given that $\lim_{n\to\infty} \frac{1}{b_n} = \frac{1}{m}$, we replaced ϵ in the usual definition of convergence with $\frac{K}{|m|}\epsilon$. This is justified in exactly the same way as we argued that ϵ can be replaced by $\frac{\epsilon}{2}$ in the proof of Theorem 4.1.1. You may find it helpful to return to the discussion of that point (see also Exercise 4.9).

Now that we've proved Theorem 4.3.1 you should go back to the sequence (S2) and convince yourself that every step can be justified to prove that the limit is 3.

4.4 Fibonacci Numbers and the Golden Section

Let's return to the Fibonacci sequence (S3). We'll do two things in this section. First we'll obtain a general formula which allows us to calculate f_n for any value of n and secondly we'll calculate the limit of the sequence (S4), i.e. we'll find $\lim_{n\to\infty} r_n = \lim_{n\to\infty} \frac{f_{n+1}}{f_n}$. Notice that we can't use algebra of limits here as $\lim_{n\to\infty} f_n = \infty$ and $\frac{\infty}{\infty}$ has no meaning. From equation (4.1.1) we know that $f_n = f_{n-1} + f_{n-2}$ and we also have the starting points $f_1 = f_2 = 1$. It will make the analysis slightly easier below if we also define $f_0 = 0$. There's nothing dodgy about this. Indeed we had zero rabbits before we started and it allows us to include the case $n = 2$ in (4.1.1). Now I'm going to rewrite the equation (4.1.1) using a different notation:

$$g_n = g_{n-1} + g_{n-2}. \tag{4.4.4}$$

What is the point of this? Well f_n is our notation for Fibonacci numbers. We know that f_n is a solution of the equation (4.4.4) but there might be other solutions that are nothing to do with Fibonacci numbers. Using the notation g_n allows us to talk generally about solutions of the equation and then we'll wind our way back to Fibonacci numbers later on.

Equation (4.4.4) is an example of a *difference equation*. It's not difficult to solve equations of this kind. We need to find a candidate solution. This is something you do by trial and error and it turns out to be sensible to experiment with a trial solution $g_n = r^n$ where $r > 0$. At this stage you should be aware that I am not saying that r^n really is a/the solution to (4.4.4). We'll just pretend that it is and then see what happens. When we substitute r^n into (4.1.1) we obtain the equation:

$$r^n = r^{n-1} + r^{n-2}.$$

This looks pleasant but perhaps hard to solve? In fact it's easy – since $r > 0$ we can divide both sides by the common factor r^{n-2} to get the quadratic equation

$$r^2 = r + 1$$

which we can solve by using the famous 'quadratic equation formula',[9] to find that the equation has two solutions:

$$r = \frac{1 \pm \sqrt{5}}{2}. \tag{4.4.5}$$

These are the only values of r for which $g_n = r^n$ solves our equation (4.4.4). The largest of these is the *golden section* $\frac{1+\sqrt{5}}{2}$ which is often denoted by the Greek letter ϕ (pronounced 'phi'). You may recall that we briefly mentioned this in our list of famous irrational numbers in Section 2.2. The other solution is the negative number $\frac{1-\sqrt{5}}{2}$ and you can easily check that

$$1 - \phi = -\frac{1}{\phi} = \frac{1 - \sqrt{5}}{2}.$$

It follows that ϕ^n and $(1 - \phi)^n$ are both solutions of the equation (4.4.4). But in what sense do we get the Fibonacci numbers from these? We have to be a little more careful. First of all you can check that $A\phi^n$ is also a solution of (4.4.4) – just multiply both sides of the equation by A. Similarly $B(1 - \phi)^n$ is a solution for any real number B. Finally we can add these together to see that $A\phi^n + B(1 - \phi)^n$ is also a solution. Although we won't give a proof here, there are no other solutions and we call

$$g_n = A\phi^n + B(1 - \phi)^n, \tag{4.4.6}$$

the *general solution*. A and B are free to take on any values that we like – but not if we want to get the Fibonacci numbers. So from now on we are going to find conditions for which $g_n = f_n$. In this case we know that $f_0 = 0$ and $f_1 = 1$ and this gives constraints on the values of A and B. To be consistent with the first of these, we put $n = 0$ and then

$$0 = A + B, \quad \text{i.e.} \quad B = -A.$$

For consistency with the second constraint, we put $n = 1$ in (4.4.6) and get

$$1 = A\phi - A(1 - \phi) = (2\phi - 1)A,$$

---

[9] The solutions to $ar^2 + br + c = 0$ are $r = \frac{-b \pm \sqrt{b^2 - 4ac}}{2a}$. In our case $a = 1, b = c = -1$.

and so

$$A = \frac{1}{2\phi - 1} = \frac{1}{\sqrt{5}}.$$

This tells us that the formula for the nth Fibonacci number is

$$f_n = \frac{1}{\sqrt{5}}[\phi^n - (1 - \phi)^n]. \tag{4.4.7}$$

Isn't that beautiful? The wonderful thing about this formula is that the combination of irrational numbers on the right-hand side always produces Fibonacci numbers which are natural numbers. You should check this for yourself in the cases $n = 2, 3, 4$.

Now we'll calculate the limit of the sequence (S4). By (4.4.7) we have

$$r_n = \frac{f_{n+1}}{f_n} = \frac{\phi^{n+1} - (1 - \phi)^{n+1}}{\phi^n - (1 - \phi)^n}.$$

We can divide top and bottom of this fraction by ϕ^n to get

$$r_n = \frac{\phi - (1 - \phi)\left(\frac{1-\phi}{\phi}\right)^n}{1 - \left(\frac{1-\phi}{\phi}\right)^n}.$$

Since $\phi > |1 - \phi|$ (why?) we have $\left|\frac{1-\phi}{\phi}\right| < 1$ and so by the results of Example 4.2, $\lim_{n \to \infty} \left|\frac{1-\phi}{\phi}\right|^n = 0$. Then by the algebra of limits (Theorem 4.3.1):

$$\lim_{n \to \infty} r_n = \phi,$$

and so we've proved that the limit of the ratio of successive Fibonacci numbers is the golden section – a result that was first suggested by the renowned astronomer Johannes Kepler (1571–1630).

Why is ϕ such an important number? It expresses a very natural relationship of great beauty which has been much exploited in geometry and architecture. Consider a section of a straight line that is divided into two parts and choose units so that the length of the smaller part has length 1. Call x the length of the larger part so that the section as a whole has length $1 + x$, as shown in Figure 4.3. To obtain the desired proportion we require x to be such that the ratio of the smaller

1   x

Figure 4.3. The golden section.

to the larger is precisely equal to the ratio of the larger to the whole, i.e.

$$\frac{1}{x} = \frac{x}{1+x}.$$

When we cross-multiply we get the quadratic equation $x^2 - x - 1 = 0$ and as we've already seen, the unique solution with $x > 0$ is $x = \phi$.

4.5 Exercises for Chapter 4

1. Consider the sequence $a_n = 1 + \frac{3}{n}$. Can you guess the limit l of this sequence?

   (a) Verify that your guess is feasible by finding a natural number n_0, for each of the following given values of ϵ such that $n > n_0 \Rightarrow |a_n - l| < \epsilon$

   (i) $\epsilon = 0.1$ (ii) $\epsilon = 0.01$ (iii) $\epsilon = 0.001$ (iv) $\epsilon = 0.0001$ (v) $\epsilon = 10^{-10}$

   (b) Give a rigorous proof that a_n converges to l.

2. Use the definition of convergence to find the limits of the sequences whose nth terms are as follows

   (a) $1 - \frac{1}{n}$ (b) $\frac{3}{n}$ (c) $\frac{1}{n^2}$ (d) $\frac{1}{\sqrt{n}}$.

   In each case you should proceed by finding the value of n_0 which ensures that 'closeness' is satisfied for any given ϵ.

3. Write down a formula for the general term of a sequence (a_n) so that a_1, a_2, a_3, a_4 and a_5 are precisely $1, \frac{2}{3}, \frac{3}{5}, \frac{4}{7}, \frac{5}{9}$ and use the definition of limit to show that the sequence converges to $\frac{1}{2}$.

4. Show that if (x_n) converges to x then $(|x_n|)$ converges to $|x|$. Is the converse true? If so, give a proof and if not, present a counter-example.

5. If $a_n \to 0$ as $n \to \infty$ and $0 \le b_n \le a_n$ for all n, show that $\lim_{n\to\infty} b_n = 0$.

6. Use the algebra of limits to find the limits of the following sequences:

   (a) $\left(2 - \frac{1}{n}\right)\left(3 + \frac{1}{n}\right)$ (b) $\left(1 + \frac{1}{\sqrt{n}}\right)^2$ (c) $\frac{2n+3}{5n+9}$

   (d) $\frac{n^2+1}{2n^2-n+2}$ (e) $\sqrt{n+1} - \sqrt{n}$

7. The following were all written down in an examination in answer to the question, 'What is the definition of a sequence (x_n) converging to a limit x?' Say what is wrong, if anything, with each of them. (a) For some $\epsilon > 0$ there is an N such

that $|x_n - x| < \epsilon$ for $n > N$. (b) Where $\epsilon > 0$, for some natural number N where $n > N$, $|x_n - x| < \epsilon$. (c) For every positive number ϵ there is a term in the sequence after which all the following terms are within ϵ of x. (d) For any $\epsilon > 0$ there is some $n > N$ such that $|x_n - x| < \epsilon$. (e) For some $\epsilon > 0$ there is a natural number $N < n$ such that $|x_n - x| < \epsilon$ for all n past a certain point.

8. The purpose of this question is to show that the order of the words in the definition of convergence is critical. A sequence (x_n) is defined to be *ridiculously convergent* to x (this is just made up for this question) if there exists a natural number N such that for every $\epsilon > 0$ we have $|x_n - x| < \epsilon$ whenever $n > N$.

    (a) Comment on the difference between 'ridiculous convergence' and 'convergence' (in the usual sense).
    (b) Show that the sequence $(\frac{1}{n})$ is not ridiculously-convergent to 0.

9. Suppose that the sequence (x_n) converges to x. Let $C > 0$ be a fixed positive constant. Show that for any $\epsilon > 0$ there is a natural number N such that $|x_n - x| < C\epsilon$ whenever $n > N$.

10. Show that if (x_n) is a sequence converging to x and that $x_n < a$ for all n then $x \le a$. [Hint: Use the result of Theorem 4.1.2.]

11. The 'sandwich rule' says that if (a_n), (b_n) and (c_n) are three sequences for which $a_n \le b_n \le c_n$ for all n and where (a_n) and (c_n) both converge to the same limit l, then (b_n) also converges to l. Prove this result. [Hint: Use the result of Exercise 4.5.]

12. Use the sandwich rule to find the limit of the sequence whose nth term is $\frac{n - \cos(n)}{n}$.

13. Consider a positive sequence (x_n), i.e. one for which each $x_n > 0$, and assume that the sequence converges to a positive limit. Show that $\lim_{n \to \infty} \frac{x_{n+1}}{x_n} = 1$. Give examples, one in each case, of a convergent positive sequence (x_n) for which the sequence whose nth term is $\frac{x_{n+1}}{x_n}$ (i) converges to zero, (ii) converges to a half, (iii) diverges (trickier).

14. (a) Let $r > 1$ and consider the sequence $(r^{\frac{1}{n}})$. Prove that it converges to 1. [Hint: Write $r^{\frac{1}{n}} = 1 + c_n$ where $c_n > 0$ and use Bernoulli's inequality from Exercise 3.7 to show that $\lim_{n \to \infty} c_n = 0$.]
    (b) Show that $\lim_{n \to \infty} r^{\frac{1}{n}} = 1$ when $0 < r < 1$. [Hint: Write $r = \frac{1}{5}$.]
    (c) Prove that $\lim_{n \to \infty} n^{\frac{1}{n}} = 1$. [Hint: Write $\sqrt{n}^{\frac{1}{n}} = 1 + c_n$.]

15. This question deals with the important notion of *subsequence*. Let (a_n) be an arbitrary sequence. To get a subsequence of (a_n) we first take any increasing sequence of natural numbers: $n_1 < n_2 < n_3 < \cdots < n_r < \cdots$ We then form the sequence $b_r = (a_{n_r})$, so this sequence begins $a_{n_1}, a_{n_2}, a_{n_3}$. Then (b_r) is called a *subsequence* of (a_n), e.g. we obtain a subsequence of $(\frac{1}{n})$ by taking every fifth

term to get $\frac{1}{5}, \frac{1}{10}, \frac{1}{15}, \ldots$ Now suppose that (a_n) converges to a limit l. Show that every subsequence of (a_n) also converges to l.

16. Find two convergent subsequences of the sequence whose nth term is $(-1)^n$. [In the exercises for Chapter 5, we will prove the *Bolzano–Weierstrass theorem* which states that every bounded sequence has at least one convergent subsequence.]

5

Bounds for Glory

Among the small there is no smallest, but always something smaller.
Anaxagoras quoted in *Philosophy of Mathematics and Natural Science*, Hermann Weyl

The limit is the hero of this book. But no great hero can achieve their destiny alone. They need companions who can help them along the way. So it is with the limit. In this chapter we'll meet its two helpers – the twin concepts of the 'sup' and the 'inf'.

We've already spent some time with bounded sequences in Section 4.2 and we've seen that every convergent sequence is bounded but that bounded sequences may not necessarily be convergent. In this chapter, we'll go into the subject of bounded sequences more deeply. To start us off let's consider the sequence (a_n) whose nth term is $a_n = \frac{1}{n} + (-1)^n$; so it is the sum of two sequences, one of which is convergent while the other diverges. The sequence begins

$$0, \frac{3}{2}, -\frac{2}{3}, \frac{5}{4}, -\frac{4}{5}, \frac{7}{6}, -\frac{6}{7}, \ldots .$$

There is a pattern here and you can see that the even terms are always of the form $a_{2n} = \frac{2n+1}{2n}$, while the odd ones can be written $a_{2n-1} = \frac{2-2n}{2n-1}$. The sequence doesn't converge, indeed it oscillates finitely. In fact it never gets larger in magnitude than the second term $\frac{3}{2}$. To see this we observe that

$$a_{2n} = \frac{2n+1}{2n} = 1 + \frac{1}{2n} \leq 1 + \frac{1}{2} = \frac{3}{2}, \quad \text{and} \quad |a_{2n-1}| = \frac{2n-2}{2n-1} \leq 1 < \frac{3}{2},$$

so we have $|a_n| < \frac{3}{2}$ for all n, and so $-\frac{3}{2} < a_n \le \frac{3}{2}$. In fact we can refine these bounds. Notice that we also have

$$a_{2n-1} = \frac{2 - 2n}{2n - 1} > \frac{1 - 2n}{2n - 1} = -1,$$

so we obtain

$$-1 < a_n < \frac{3}{2}.$$

We call $\frac{3}{2}$ an *upper bound* for the sequence (a_n) and -1 a *lower bound*. Upper and lower bounds give us additional tools for describing the way in which sequences behave.

More generally we say that an arbitrary sequence (a_n) is *bounded above* if there exists a real number L such that $a_n \le L$ for all n and L is then called an upper bound for the sequence. A sequence is *bounded below* if there exists a real number M such that $a_n \ge M$ for all n and M is called a lower bound.

Three points to note.

1. Upper and lower bounds are not unique. For example if we return to the sequence with $a_n = \frac{1}{n} + (-1)^n$ then e.g. 5 is also an upper bound and -21.6 is a lower bound.

2. A sequence may be bounded above but fail to be bounded below and vice versa. For example the sequence of natural numbers with $a_n = n$ is bounded below but not bounded above. Similarly the sequence $a_n = -n$ is bounded above but not bounded below.

3. A sequence is bounded if and only if it is both bounded above and bounded below (see Exercise 5.1).

As upper and lower bounds are not unique we might enquire which of these is the 'best' for a given sequence. By 'best' here we mean the smallest (mathematicians prefer to say 'least') upper bound for a sequence that is bounded above and the greatest lower bound for one that is bounded below. If the sequence only took a finite number of distinct values then we would be looking for the maximum and minimum (respectively) as described in Section 3.4. When a sequence takes an infinite number of different values there is no reason why the maximum and minimum should exist. For example consider the sequence with general term $a_n = \frac{1}{n}$. Later on we'll prove that its greatest lower bound is 0 but we've already pointed out that there is no number N for which $\frac{1}{N} = 0$, so 0 cannot be the minimum.

It's a deep and subtle property of the real numbers that any sequence that is bounded above has a least upper bound and (consequently, as will follow from Theorem 5.1.2 – see Exercise 5.5) any sequence that is bounded below has a greatest lower bound. For now we'll just assume this but we will come back to look at this result in greater detail in Chapter 11.

Now some notation. In old books on analysis the greatest lower bound and least upper bound often used to be denoted g.l.b. and l.u.b. (respectively), but for a long time now mathematicians have preferred a different terminology which has its root in Latin. So the greatest lower bound is referred to as the *infimum* and shortened to inf while the least upper bound is called the *supremum* and denoted by sup. These come from the same Latin root as 'inferior' and 'superior'.[1]

Now it's time for a formal definition. Suppose that (a_n) is a sequence that is bounded above. We define its supremum $\sup(a_n)$ to be the unique real number that satisfies

$$a_n \leq \sup(a_n) \leq L,$$

for all n where L is any upper bound for (a_n). Similarly if (a_n) is a sequence that is bounded below, its infimum, $\inf(a_n)$, is defined to be the unique real number that satisfies

$$K \leq \inf(a_n) \leq a_n,$$

for all n where K is any lower bound for (a_n).

If there exists a natural number N such that $\sup(a_n) = a_N$ we say that the supremum is *attained*. A similar definition holds for the infimum.

As an example, we see that $\sup\left(\frac{1}{n}\right) = 1$ as $\frac{1}{n} < 1$ for all $n > 1$. As $a_1 = 1$ we see that the supremum is attained in this case. On the other hand 0 is a lower bound and suppose that α is a larger lower bound. Then $0 < \alpha \leq \frac{1}{n}$ for all n and so $n \leq \frac{1}{\alpha}$ for all n which is impossible, so we have a contradiction and can assert that $\inf\left(\frac{1}{n}\right) = 0$. So in this case the infimum is not attained (but it does coincide with the limit of the sequence).

We can see from this definition that sup coincides with max and inf is precisely min if (a_n) has only finitely many values. But on the other hand, these are more subtle concepts that are not so far removed from limits as the following theorem shows.

Theorem 5.1.1.

1. If the sequence (a_n) is bounded above then given any $\epsilon > 0$ there exists a natural number n such that

$$a_n > \sup(a_n) - \epsilon. \tag{5.1.1}$$

2. If the sequence (a_n) is bounded below then given any $\epsilon > 0$ there exists a natural number n such that

$$a_n < \inf(a_n) + \epsilon. \tag{5.1.2}$$

---

[1] Try typing supremum and infimum into http://ablemedia.com/ctcweb/showcase/wordsonline.html

Proof. We'll only prove (1) as the proof for (2) is similar and we'll employ a proof by contradiction. So suppose the result is false and no such n exists. Then $a_n < \sup(a_n) - \epsilon$ for all n. But then $\sup(a_n) - \epsilon$ is an upper bound for (a_n) that is smaller than $\sup(a_n)$. But $\sup(a_n)$ is the smallest upper bound (by definition) and we have our contradiction. □

Although I said that Theorem 5.1.1 has some similarities with the definition of the limit, I hope you'll see that it presents much weaker statements. For example, the first of these tells us that as we go along the sequence we must be able to find a sufficiently large n such that a_n gets arbitrarily close to $\sup(a_n)$, but the direction of closeness is only from one side and we say nothing about what happens for larger values of n. Later in this chapter we will see that sups and infs can sometimes be limits but we will need to impose more structure on the type of sequence we consider.

There is a natural symmetry between the concepts of sup and inf – indeed to get from one to the other we just replace 'greatest' by 'least' and 'upper' by 'lower'. In fact, there is a sense in which we only need one of these concepts – inf is really a sup in disguise as the following theorem proves.

Theorem 5.1.2. If (a_n) is a bounded sequence then

$$\inf(a_n) = -\sup(-a_n).$$

Proof. As $-a_n \le \sup(-a_n)$ for all n, by (L4) $-\sup(-a_n) \le a_n$ for all n and so $-\sup(-a_n)$ is a lower bound for our sequence. To show that it is really the inf we'll assume that it isn't and try to obtain a contradiction. So assume that there exists a real number β such that

$$-\sup(-a_n) < \beta \le a_n,$$

for all n. But using (L4) again we get

$$-a_n \le -\beta < \sup(-a_n),$$

and so $-\beta$ is a smaller upper bound for the sequence $(-a_n)$ than its own supremum – which is the contradiction we were looking for. □

This is fairly typical of how results about sups and infs are proved. If you've guessed that a number α might be $\sup(a_n)$ you should firstly show that α really is an upper bound for the sequence and secondly assume that it isn't the sup and try to find a contradiction. The next result we'll prove is a fairly simple one but it is quite useful.

Lemma 5.1.

1. If the sequence (a_n) is bounded and is nonnegative, i.e. each $a_n \ge 0$ then

$$\sup(a_n) \ge \inf(a_n) \ge 0.$$

2. If (b_n) and (c_n) are bounded sequences with $b_n \geq c_n$ for all n then

$$\inf(b_n) \geq \inf(c_n) \quad \text{and} \quad \sup(b_n) \geq \sup(c_n).$$

Proof.

1. Clearly 0 is a lower bound for (a_n) so we must have $\inf(a_n) \geq 0$. But $\sup(a_n) \geq \inf(a_n)$ by definition and that's all we need.

2. We have $\inf(c_n) \leq c_n \leq b_n$ for all n and so $\inf(c_n)$ is a lower bound for (b_n). In this case it cannot exceed the greatest lower bound and so $\inf(c_n) \leq \inf(b_n)$. The result for the sup is proved similarly. $\square$

Suppose that the sequences (a_n) and (b_n) are both bounded above then so is the sequence $(a_n + b_n)$. Indeed if K is an upper bound for (a_n) and L is an upper bound for (b_n) you can check easily that $K + L$ is an upper bound for $(a_n + b_n)$. You may then guess that we might have '$\sup(a_n + b_n) = \sup(a_n) + \sup(b_n)$'. As is so often the case in this subject, we have to be more careful. Equality doesn't hold as we can see by taking $a_n = 1 + \frac{1}{n}$ and $b_n = -\frac{1}{n}$. Then $\sup(a_n + b_n) = 1$, $\sup(a_n) = 2$ and $\sup(b_n) = 0$. What can be proved is a weaker but useful result.

Theorem 5.1.3. If (a_n) and (b_n) are bounded above then

$$\sup(a_n + b_n) \leq \sup(a_n) + \sup(b_n).$$

Proof. Since $a_n \leq \sup(a_n)$ and $b_n \leq \sup(b_n)$ we have

$$a_n + b_n \leq \sup(a_n) + \sup(b_n),$$

for all n. So $\sup(a_n) + \sup(b_n)$ is an upper bound for $(a_n + b_n)$. Then it cannot be smaller than the least upper bound and that gives our result. $\square$

You might expect that a similar result to Theorem 5.1.3 holds for the inf. It does, but you have to be careful as it goes the other way around. The result is that if (a_n) and (b_n) are both bounded below then so is $(a_n + b_n)$ and

$$\inf(a_n) + \inf(b_n) \geq \inf(a_n + b_n).$$

You can prove this for yourself either by imitating the proof of Theorem 5.1.3 or by combining the result of that theorem with that of Theorem 5.1.2 to turn infs into sups (see Exercise 5.4).

5.2 Monotone Sequences

In this section we'll focus on the question I posed before – when can a sup or an inf be a genuine limit? The answer to this is when a sequence is a monotone one.

To be precise we will say that a sequence (a_n) is *monotonic increasing* if $a_{n+1} \geq a_n$ for all n and is *monotonic decreasing* if $a_{n+1} \leq a_n$ for all n. Finally we say that a sequence is *monotone* if it is either monotonic increasing or decreasing. An example of a monotonic decreasing sequence is $a_n = \frac{1}{n}$ while $a_n = 1 - \frac{1}{n}$ is monotonic increasing. To prove the first is straightforward. To prove the second you can either show that we always have $a_{n+1} - a_n \geq 0$ by doing some algebra, or use the (fairly obvious) fact that a sequence (a_n) is monotonic increasing if and only if $(-a_n)$ is monotonic decreasing.

Now it certainly isn't true that every monotone sequence converges, e.g. think of $a_n = n$. But suppose a sequence is monotonic increasing and bounded above. Then on the one hand we are told that our sequence is steadily increasing in value, but on the other hand, we have imposed a ceiling on it that it cannot exceed. So where can it go to except to the ceiling? The next result puts this intuition into precise mathematical form.

Theorem 5.2.1.

1. If the sequence (a_n) is bounded above and monotonic increasing then it is convergent and $\lim_{n \to \infty} a_n = \sup(a_n)$.

2. If the sequence (a_n) is bounded below and monotonic decreasing then it is convergent and $\lim_{n \to \infty} a_n = \inf(a_n)$.

Proof.

1. Since (a_n) is bounded above we know that $\alpha = \sup(a_n)$ exists. We also know from Theorem 5.1.1 that given any $\epsilon > 0$ there exists n_0 such that $a_{n_0} > \alpha - \epsilon$. But (a_n) is monotonic increasing so for all $n \geq n_0$ we have $a_n \geq a_{n_0+1} \geq a_{n_0} > \alpha - \epsilon$. This tells us by simple algebra that

$$\alpha - a_n < \epsilon, \quad \text{for all} \quad n \geq n_0.$$

But $\alpha > a_n$ for all n and so $|\alpha - a_n| = \alpha - a_n$. Then we've satisfied the conditions for convergence as described in the definition of the concept and can assert that $\lim_{n \to \infty} a_n = \alpha$.

2. This is a good opportunity to test your understanding by doing it yourself. There are two approaches. The first is to imitate the proof we've just given by using the second part of Theorem 5.1.1. The second approach which is perhaps a little slicker is to derive the result as a corollary to (1) by using the fact that a sequence (a_n) is monotonic decreasing and bounded below if and only if the sequence $(-a_n)$ is monotonic increasing and bounded above and then applying (1) and Theorem 5.1.2. □

You might think that the next result is too obvious to need a proof – but I hope you've seen enough by now to appreciate that the obvious isn't always true. In mathematics, everything must be proved logically.

Corollary 5.2.1.

1. If the sequence (a_n) is monotonic increasing then either it converges or it diverges to $+\infty$.

2. If the sequence (a_n) is monotonic decreasing then either it converges or it diverges to $-\infty$.

Proof. We'll only prove (2) as (1) is so similar. Suppose that (a_n) is monotonic decreasing. Then either it is bounded below or it isn't. If it is bounded below then it converges by Theorem 5.2.1 (2). If it isn't bounded below then given any $K < 0$ we can find a natural number n_0 such that $a_{n_0} < K$ for otherwise K would be a lower bound. But then since the sequence is monotonic decreasing we have $a_n < K$ for all $n \geq n_0$ and so the sequence diverges to $-\infty$. $\square$

5.3 An Old Friend Returns

To get a feel for how to use Theorem 5.2.1 we should do an example.[2] We'll construct a sequence (a_n) by *recursion* so that a_{n+1} is not given explicitly by a known formula but implicitly through the value of a_n. This doesn't work unless we have a starting point and so we define (a_n) by:

$$a_1 = 1 \quad \text{and} \quad a_{n+1} = \sqrt{1 + a_n} \quad \text{for} \quad n \geq 1. \qquad (5.3.3)$$

Let's calculate the first few terms. We have

$$a_2 = \sqrt{1 + 1} = 1.4142136\ldots$$

$$a_3 = \sqrt{1 + \sqrt{2}} = 1.553774\ldots$$

$$a_4 = \sqrt{1 + \sqrt{1 + \sqrt{2}}} = 1.5980532\ldots$$

$$a_5 = \sqrt{1 + \sqrt{1 + \sqrt{1 + \sqrt{2}}}} = 1.6118478\ldots$$

$$a_6 = \sqrt{1 + \sqrt{1 + \sqrt{1 + \sqrt{1 + \sqrt{2}}}}} = 1.6161212\ldots$$

It certainly looks like (a_n) is increasing and bounded above. How do we prove this? Let's look at the bounded problem first.

Bounded. From the calculations we've done it certainly looks like 2 will be an upper bound. There's no good reason why it should be the sup but finding that isn't our concern ... yet. Let's use a proof by contradiction and suppose that there

---

[2] You may find it helpful to attempt Exercise 5.6 before you read the rest of this section.

exists a number N such that $a_n \leq 2$, for all $1 \leq n \leq N$, but $a_{N+1} > 2$. From our calculations we know that if it exists then $N > 5$. Now by (5.3.3), $\sqrt{1 + a_N} > 2$ and squaring this (remember (L6)) yields

$$1 + a_N > 4,$$

i.e. $a_N > 3$. That's a contradiction and so we can assert that our sequence really is bounded above.

Monotone. Squaring the general recursive formula in (5.3.3) we get for all $n \geq 1$,

$$a_{n+1}^2 = 1 + a_n,$$

and for all $n \geq 2$,

$$a_n^2 = 1 + a_{n-1}.$$

Subtracting the second equation from the first yields

$$a_{n+1}^2 - a_n^2 = a_n - a_{n-1},$$

i.e. $(a_{n+1} + a_n)(a_{n+1} - a_n) = a_n - a_{n-1},$

and so,[3]

$$a_{n+1} - a_n = \frac{a_n - a_{n-1}}{a_{n+1} + a_n}.$$

Working backwards we get

$$a_n - a_{n-1} = \frac{a_{n-1} - a_{n-2}}{a_n + a_{n-1}},$$

and continuing in this manner we eventually get to

$$a_3 - a_2 = \frac{a_2 - a_1}{a_3 + a_2}.$$

Combining all of these together we find that

$$a_{n+1} - a_n = \frac{a_2 - a_1}{(a_{n+1} + a_n)(a_n + a_{n-1}) \cdots (a_3 + a_2)}$$

$$= \frac{\sqrt{2} - 1}{(a_{n+1} + a_n)(a_n + a_{n-1}) \cdots (a_3 + a_2)} > 0,$$

as $\sqrt{2} > 1$ and the bottom line of the fraction is a positive number. This shows that $a_{n+1} \geq a_n$ for all n and so (a_n) is monotonic increasing as we wanted.

---

[3] You can use a similar argument to the one we have given to prove that the sequence is bounded above to show $a_n > 1$ for all n and so the division below is justified.

Limit. As the sequence is bounded above and monotonic increasing we know that it converges by Theorem 5.3.3. Let $l = \sup(a_n) = \lim_{n\to\infty} a_n$. To find l we'll first square both sides of (5.3.3) to get

$$a_{n+1}^2 = 1 + a_n,$$

and then take limits of both sides

$$\lim_{n\to\infty} a_{n+1}^2 = \lim_{n\to\infty} (1 + a_n).$$

Now apply the algebra of limits (Theorem 4.3.1) and we obtain a quadratic equation in l:

$$l^2 = l + 1,$$

$$\text{i.e.} \quad l^2 - l - 1 = 0.$$

We've met this equation before in Section 4.4 where we learned that it has two solutions – the golden section ϕ and $1 - \phi$. In our case, since the limit is the sup and every term of the sequence is a positive number, we must have $l > 0$ and so $l = \phi$. So we find the golden section appearing in another guise – as the limit of the sequence defined by (5.3.3).

5.4 Finding Square Roots

We've already pointed out that one of our goals is to be able to find (at least in principle) any irrational number as the limit of a sequence of rationals. In the last section we saw how to do this for the golden section. We are pretty far from being able to do this in general, but we can at least look at the square roots of prime numbers which we showed were irrational back in Chapter 2. Our aim is to find $\sqrt{p}$ where p is any prime number and to do this we again set up a recursive sequence which this time is defined by

$$a_1^2 \geq p \quad \text{and} \quad a_{n+1} = \frac{1}{2}\left(a_n + \frac{p}{a_n}\right) \quad \text{for} \quad n \geq 1, \tag{5.4.4}$$

where I also insist that a_1 is a rational number. It seems strange to define a_1 by an inequality but here all I am saying is that I have perfect freedom to choose a_1 however I please, provided that its square exceeds the number p whose square root we seek. So for example if $p = 5$ we could take $a_1 = 2.5$ or $a_1 = 3$ but not $a_1 = 2$. If this bothers you, then there is no harm done in just choosing $a_1 = p$. I will return to this discussion of 'starting points' at the very end of the section. Note that each a_n is rational. For suppose that $a_1, a_2, \ldots, a_{N-1}$ are rational but

a_N isn't. We know $N > 1$ as a_1 is rational. Now as a_{N-1} is rational we can write $a_{N-1} = \frac{x}{y}$ where x and y are natural numbers. Then

$$a_N = \frac{1}{2}\left(\frac{x}{y} + \frac{yp}{x}\right),$$

is also a rational number and that's the contradiction we were looking for. In the argument that I just gave, I assumed that a_n was always a positive number. This really should have been proved first but I'll leave that step to the reader. It isn't hard and works by a similar argument to the one I've just given.

Before we go any further, let's look at a special case. We'll take $p = 5$. Then systematically applying (5.4.4) (and choosing $a_1 = 5$) we find $a_2 = 3$, $a_3 = 2.33333\ldots, x_4 = 2.2380952\ldots, x_5 = 2.2360689\ldots, x_6 = 2.236068$. I'll stop here as after five uses of (5.4.4) or *iterations*, we have found $\sqrt{5}$ correct to six decimal places. This is pretty impressive.

Returning to the general case, we'd like to show that $\lim_{n\to\infty} a_n = \sqrt{p}$. The obvious strategy is to use Theorem 5.2.1 by showing that the sequence (a_n) is bounded below and monotonic decreasing.

Now notice that if we know that (a_n) converges then it's limit really is $\sqrt{p}$. To see that write $\alpha = \lim_{n\to\infty} a_n$ and use algebra of limits in (5.4.4) to deduce that

$$\alpha = \frac{1}{2}\left(\alpha + \frac{p}{\alpha}\right),$$

and a little algebra yields $\alpha^2 = p$ and so $\alpha = \sqrt{p}$ (we can't have $\alpha = -\sqrt{p}$ by Theorem 4.1.2).

Now in Exercise 3.10 you should have verified the inequality

$$p \leq \frac{1}{4}\left(y + \frac{p}{y}\right)^2 \leq y^2 \tag{5.4.5}$$

whenever $y^2 \geq p$. Since $a_1^2 \geq p$ (and now you know why I insisted on this) if we put $y = a_1$ then we see from (5.4.5) that we also have $a_2^2 \geq p$. Then use (5.4.5) again with $y = a_2$ to get $a_3 \geq p$. Now suppose that $a_r^2 \geq p$ for all $1, 2, \ldots, n$ but that $a_{n+1}^2 < p$. Then (5.4.5) yields $p \leq \frac{1}{4}\left(a_n + \frac{p}{a_n}\right)^2 = a_{n+1}^2$ and we have deduced a contradiction. Hence we see that we must have $a_n^2 \geq p$ for all n and so the sequence (a_n) is bounded below and $\sqrt{p}$ is a lower bound.

We'll now prove that (a_n) is monotonic decreasing. We compute

$$a_n - a_{n+1} = a_n - \frac{a_n}{2} - \frac{p}{2a_n}$$

$$= \frac{1}{2}\left(a_n - \frac{p}{a_n}\right)$$

$$= \frac{1}{2a_n}(a_n^2 - p) \geq 0,$$

as we have just shown that $a_n^2 \geq p$ for all n.

The method we've just used for finding square roots of primes can just as easily be used to find the square root of any positive real number. There is another quite straightforward way to obtain square roots without using Theorem 5.2.1 and this is called the 'Newton-Raphson method'.[4] It relies on calculus and so is outside the scope of this book. In fact when we use this, we can forget about trying to make the sequence monotonic decreasing from the outset which I did here by making the choice $a_1 \geq p$. You get quicker convergence if you start at a point which is a rough guess at $\sqrt{p}$ so for the case $p = 5$, since $2^2 = 4$ and $3^2 = 9$, you might choose $a_1 = 2.2$ or 2.3.

5.5 Exercises for Chapter 5

1. Prove that a sequence (a_n) is bounded (i.e. there exists $K > 0$ such that $|a_n| \leq K$ for all n) if and only if it is bounded above and bounded below.

2. For each of the following sequences say whether it is (i) bounded above or below, (ii) monotone increasing or decreasing, (iii) convergent. In the case of (i), write down an upper or lower bound (as appropriate) and try to guess the supremum and/or infimum (or even better, establish these by a proof) and in (iii), write down the limit when this exists.

   (a) $3 - \dfrac{5}{n!}$ (b) $\dfrac{1}{1 - \frac{1}{n+1}}$ (c) $\dfrac{n}{1 - \frac{1}{n+1}}$ d) $(-1)^n \dfrac{1}{n}$ (e) $\dfrac{1}{n}(1 - (-1)^n)$

3. If (a_n) and (b_n) are sequences which are bounded above, show that

   (a)
   $$\sup(a_n b_n) \leq \sup(a_n)\sup(b_n),$$

   whenever each $a_n, b_n \geq 0$.
   Give a counter-example to show that equality does not hold in general.
   Show further that

   (b)
   $$\sup(|a_n + b_n|) \leq \sup(|a_n|) + \sup(|b_n|),$$

   (c)
   $$|\sup(a_n)| \leq \sup(|a_n|).$$

4. Suppose that the sequences (a_n) and (b_n) are bounded below. Deduce that
   $$\inf(a_n) + \inf(b_n) \geq \inf(a_n + b_n).$$

[4] See e.g. http://en.wikipedia.org/wiki/Newton's_method. If you know what the method is then to find square roots of primes you need to apply it to $f(x) = x^2 - p$.

Also find and prove an analogue to part (a) of the previous question. Give counter-examples to show that equality doesn't always hold in both cases.

5. Assume the completeness property of the real numbers (see Chapter 11), i.e. that every sequence (x_n) that is bounded above has a least upper bound. Prove that every sequence that is bounded below has a greatest lower bound. [Hint: Use Theorem 5.1.2.]

6. Let $x_1 = 2.5$ and $x_{n+1} = \frac{1}{5}(x_n^2 + 6)$ for $n > 1$.

   (a) Show that each $2 \le x_n \le 3$. (Hint: Try a proof by contradiction.)
   (b) Show that $x_{n+1} - x_n = \frac{1}{5}(x_n - 2)(x_n - 3)$.
   (c) Show that the sequence (x_n) is monotone and find its limit as $n \to \infty$.

7. Let $a \ge b > 0$. We define sequences (a_n) and (b_n) by taking a_1 and b_1 to be a and b respectively, and requiring that for $n \ge 1$ $a_{n+1} = \frac{1}{2}(a_n + b_n)$ and $b_{n+1} = \sqrt{a_n b_n}$. In other words, a_{n+1} is the arithmetic mean of a_n and b_n while b_{n+1} is their geometric mean.

   (a) Prove that $b_n \le b_{n+1} \le a_{n+1} \le a_n$ for each n.
   (b) Prove that $a_{n+1} - b_{n+1} \le \frac{1}{2}(a_n - b_n)$ for all n.
   (c) Deduce that the sequences (a_n) and (b_n) are each convergent and that they converge to the same limit. (The common limit $M(a; b) = \lim_{n \to} a_n = \lim_{n \to \infty} b_n$ is called the *arithmetic-geometric mean* of a and b. It can be given a precise form using objects called *elliptic integrals*.)

8. Show that if (a_n) is a sequence that is both monotonic increasing and also convergent to a limit l as $n \to \infty$, then (a_n) is bounded above and $l = \sup(a_n)$. What happens when (a_n) is monotonic decreasing and convergent?

9. The purpose of this question is to prove that $n^p x^n \to 0$ as $n \to \infty$ for any real number p and for any $-1 < x < 1$. Assume firstly that $0 < x < 1$, and write $a_n = n^p x^n$.(a) Show that $\lim_{n \to \infty} \frac{a_{n+1}}{a_n} = x$. (b) Deduce that $\frac{a_{n+1}}{a_n}$ is eventually less than one and so (a_n) is eventually decreasing. [Here 'eventually' means there is some N such that the statement is true for all $n > N$.] (c) Deduce that the sequence (a_n) tends to a nonnegative limit l. (d) Use part (a) with Exercise 4.13 to deduce that $l = 0$. What about the case where $-1 < x < 0$?

10. Suppose that (a_n) is a monotonic increasing sequence that has a subsequence (a_{n_k}) which converges to a limit l.

    (a) Show that $a_n \le l$ for all n.
    (b) Show that (a_n) converges to l.
    (c) What happens when 'increasing' is replaced by 'decreasing' in this question?

11. As promised at the end of Exercise 4.16 we will now prove the celebrated *Bolzano-Weierstrass theorem* which states that every bounded sequence has at least one convergent subsequence. Let (x_n) be a sequence for which $a \le x_n \le b$ for all n.

Define $a_1 = a$ and $b_1 = b$. Let $c_1 = \frac{1}{2}(b_1 - a_1)$. Either an infinite number of terms of the sequence (x_n) lie in the interval $[a_1, c]$ or they lie in $[c, b]$ (or they lie in both). Suppose for the sake of argument that they lie in $[a_1, c]$. Define $a_2 = a_1$ and $b_2 = c_1$. Define $c_2 = \frac{1}{2}(b_2 - a_2)$ and repeat the argument just given. In fact repeat this exercise indefinitely to generate two sequences of numbers (a_n) and (b_n) where

$$a_1 \leq a_2 \leq \cdots \leq a_n \leq \cdots \leq b_n \leq \cdots b_2 \leq b_1.$$

(a) Deduce that $b_n - a_n = \frac{1}{2^{n-1}}(b - a)$ for all n.
(b) Use Theorem 5.2.1 to show that (a_n) and (b_n) both converge. Use the result of (a) to verify that $\lim_{n\to\infty} a_n = \lim_{n\to\infty} b_n$.
(c) Explain how you may extract a subsequence (x_{n_r}) of (x_n) for which $a_r \leq x_{n_r} \leq b_r$ for each r. Hence show that (x_{n_r}) converges.

12. Let (x_n) be a bounded sequence and define two associated sequences as follows

$$a_n = \sup(x_m, m \geq n) \quad \text{and} \quad b_n = \inf(x_m, m \geq n)$$

(a) Show that (a_n) is monotonic decreasing, bounded below and hence convergent.
(b) Show that (b_n) is monotonic increasing and bounded above and hence convergent.

We define

$$\limsup_{n\to\infty} x_n = \lim_{n\to\infty} a_n,$$

$$\liminf_{n\to\infty} x_n = \lim_{n\to\infty} b_n.$$

Find lim sup and lim inf of the following sequences:

(i) $(-1)^n$, (ii) $\dfrac{1}{n}$ (iii) $(-1)^n(1 - \dfrac{1}{n})$

Note: lim sup and lim inf play a major role in advanced analysis. An important theorem states that a bounded sequence (x_n) converges to the limit l if and only if

$$\limsup_{n\to\infty} x_n = \liminf_{n\to\infty} x_n = l.$$

You may encounter some books in which $\limsup_{n\to\infty} x_n$ is written $\overline{\lim}_{n\to\infty} x_n$ and $\liminf_{n\to\infty} x_n$ is written $\underline{\lim}_{n\to\infty} x_n$.

You Cannot be Series

The limiting process was victorious. For the limit is an indispensable concept, whose importance is not affected by the acceptance or rejection of the infinitely small.

Philosophy of Mathematics and Natural Science, Hermann Weyl

6.1 What are Series?

Sequences are the fundamental objects in the study of limits. In this chapter we will meet a very special type of sequence whose limit (when it exists) is the best meaning we can give to the intuitive idea of an 'infinite sum of numbers'. Let's be specific. Suppose we are given a sequence (a_n). Our primary focus in this chapter will not be on the sequence (a_n) but on a related sequence which we will call (s_n). It is defined as follows:

$$s_1 = a_1$$
$$s_2 = a_1 + a_2$$
$$s_3 = a_1 + a_2 + a_3,$$

and more generally

$$s_n = a_1 + a_2 + \cdots + a_{n-1} + a_n.$$

The sequence (s_n) is called a *series* (or infinite series) and the term s_n is often called the *nth partial sum* of the series. The goal of this chapter will be to investigate when the series (s_n) has a limit. In this way we can try to give meaning to 'infinite sums' which we might write informally as

$$1 + \frac{1}{2} + \frac{1}{3} + \frac{1}{4} + \cdots, \tag{6.1.1}$$

78

$$1 - 1 + 1 - 1 + 1 - 1 + \cdots, \tag{6.1.2}$$

but beware that the meaning that we'll eventually give to expressions like this (in those cases where it is indeed possible) will not be the literal one of an infinite number of additions (what can that mean?), but as the limit of a sequence. Sequences that arise in this way turn up throughout mathematics and its applications so understanding them is very important.

6.2 The Sigma Notation

Before we can start taking limits of series we need to develop some useful notation for finite sums that will simplify our approach to expressions like (6.1.1) and (6.1.2).[1]

Let's suppose that we are given the following ten whole numbers: $a_1 = 3, a_2 = -7, a_3 = 5, a_4 = 16, a_5 = -1, a_6 = 3, a_7 = 10, a_8 = 14, a_9 = -2, a_{10} = 0$. We can easily calculate their sum

$$a_1 + a_2 + \cdots + a_{10} = 41, \tag{6.2.3}$$

but there is a more compact way of writing the left-hand side which is widely used by mathematicians and those who apply mathematics. It is called the 'sigma notation' because it utilises the Greek letter Σ (pronounced 'sigma') which is capital S in English (and S should be thought of here as standing for 'sum'). Using this notation we write the left-hand side of (6.2.3) as

$$\sum_{i=1}^{10} a_i.$$

Now if you haven't seen it before, this may appear to be a complex piece of symbolism – but don't despair. We'll unpick it slowly and we'll read bottom up. The $i = 1$ tells us that the first term in our addition is a_1, then the Σ comes into play and tells us that we must add a_2 to a_1 and then a_3 to $a_1 + a_2$ and then a_4 to $a_1 + a_2 + a_3$ and then … but when do we stop? Well go to the top of Σ and read the number 10. That tells you that a_{10} is the last number you should add, and that's it. The notation is very flexible. For example you also have

$$\sum_{i=1}^{8} a_i = a_1 + a_2 + \cdots + a_8 = 43,$$

$$\sum_{i=2}^{9} a_i = a_2 + a_3 + \cdots + a_9 = 38,$$

[1] Readers who already know about this notation may want to omit this section.

$$\sum_{i=6}^{7} a_i = a_6 + a_7 = 13.$$

The following results are easily derived and quite useful. We will use them without further comment in the sequel. If $a_1, \ldots, a_n, b_1, \ldots b_n$ and c are arbitrary real numbers then

$$\sum_{i=1}^{n}(a_i + b_i) = \sum_{i=1}^{n} a_i + \sum_{i=1}^{n} b_i,$$

$$\sum_{i=1}^{n} ca_i = c \sum_{i=1}^{n} a_i.$$

The sigma notation is particularly useful when we have a sequence (a_n) and we want to add as far as the general term a_n to obtain the nth partial sum s_n. We can now write

$$s_n = \sum_{i=1}^{n} a_i.$$

Notice that the letter i here is only playing the role of a marker in telling us where to start and stop adding. It is called a 'dummy index' and the value of s_n is unchanged if it is replaced by a different letter throughout – e.g.

$$\sum_{i=1}^{n} a_i = \sum_{j=1}^{n} a_j.$$

It's worth pointing out one special case that often confuses students. Suppose that $a_i = k$ takes the same value for all i. Then

$$\sum_{i=1}^{n} a_i = \sum_{i=1}^{n} k = \underbrace{k + k + \cdots + k}_{n \text{ times}} = nk.$$

Our main concern in this chapter is with infinite rather than finite series, but before we return to the main topic let's look at an interesting problem that (and this may be an apocryphal story) was given to one of the greatest mathematicians the world has even seen, Carl Friedrich Gauss (1777–1855),[2] when he was a schoolboy. The story goes that the teacher wanted to concentrate on some urgent task and so he asked the whole class to calculate the sum of the first 100 natural numbers so that he could work in peace while they struggled with this fiendish problem. Apparently after 5 minutes Gauss produced the correct answer 5050. How did he get this? It is speculated that he noticed the following clever pattern

---

[2] See e.g. http://202.38.126.65/navigate/math/history/Mathematicians/Gauss.html

by writing the numbers 1 to 50 in a row and then the numbers 51 to 100 in reverse order underneath:

$$\begin{array}{cccccc} 1 & 2 & 3 & 4 & \cdots & 49 & 50 \\ 100 & 99 & 98 & 87 & \cdots & 52 & 51 \end{array}$$

Now each column adds to 101 – but there are 50 columns and so the answer is $50 \times 101 = 5050$. In sigma notation, Gauss calculated $\sum_{i=1}^{100} i$. A natural generalisation is to seek a formula for $\sum_{i=1}^{n} i$, i.e. the sum of the first n natural numbers. If $n = 2m$ is an even number you can use exactly the same technique as Gauss to show that the answer is $m(2m + 1)$ or $\frac{1}{2}n(n + 1)$. The same is true when $n = 2m + 1$ is odd since (using the fact that we know the answer when n is even),

$$\sum_{i=1}^{2m+1} i = \sum_{i=1}^{2m} i + (2m + 1)$$

$$= m(2m + 1) + (2m + 1)$$

$$= (m + 1)(2m + 1)$$

$$= \frac{1}{2}n(n + 1).$$

So we've shown that for every natural number n

$$\sum_{i=1}^{n} i = \frac{1}{2}n(n + 1). \tag{6.2.4}$$

At this stage it's worth briefly considering more general finite sums which are obtained from summing all the numbers on the list $a, a + d, a + 2d, \ldots, a + (n - 1)d$. Note that there are n numbers in this list which are obtained from the *first term* a by repeatedly adding the *common difference* d. Such a list is called an *arithmetic progression* and we can find the sum S of the first n terms by using a slight variation of Gauss' technique. So we write

$$S = \quad a \quad + \quad (a + d) \quad + \quad (a + 2d) \quad + \cdots + [a + (n - 1)d]$$
$$S = [a + (n - 1)d] + [a + (n - 2)d] + [a + (n - 3)]d + \cdots + \quad a$$

Now add these two expressions, noting that each of the n columns on the right-hand side sums to $2a + (n - 1)d$ to get

$$2S = n(2a + (n - 1)d),$$

and so

$$S = n\left[a + \frac{1}{2}(n - 1)d\right].$$

If you take $a = d = 1$, you can check that you get another proof of (6.2.4).

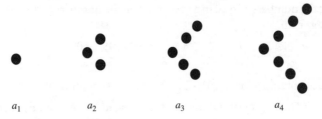

Figure 6.1. Representation of some triangular numbers.

Before we return to our main topic, I can't resist introducing the *triangular numbers*. This is the sequence (a_n) defined by $a_1 = 1, a_2 = 1 + 2 = 3$, $a_3 = 1 + 2 + 3 = 6, a_4 = 1 + 2 + 3 + 4 = 10$, etc. so the nth term is $a_n = \frac{1}{2}n(n+1)$. Figure 6.1 above should help you see why these numbers are called 'triangular'.[3]

It is a fact that if you add successive triangular numbers you always get a perfect square so e.g.

$$1 + 3 = 4 = 2^2$$

$$3 + 6 = 9 = 3^2$$

$$6 + 10 = 16 = 4^2.$$

It is easy to prove that this holds in general by using (6.2.4). Indeed we have that the sum of the nth and $(n + 1)$th triangular numbers is

$$a_n + a_{n+1} = \frac{1}{2}n(n+1) + \frac{1}{2}(n+1)(n+2)$$

$$= \frac{1}{2}(n+1)(n+n+2)$$

$$= (n+1)^2.$$

6.3 Convergence of Series

Let's return to the main topic of this chapter. We are given a sequence (a_n) and we form the associated sequence of partial sums (s_n) where $s_n = \sum_{i=1}^{n} a_i$. Now suppose that (s_n) converges to some real number l in the sense of Chapter 4, i.e. for any $\epsilon > 0$ there exists a natural number n_0 such that whenever $n > n_0$ we have $|s_n - l| < \epsilon$. In this case we call l the *sum of the series*. In some ways this is a bad name as l is not a sum in the usual sense of the word, it is a limit of sums,

---

[3] See also http://en.wikipedia.org/wiki/Triangular_number.

but this is standard terminology and we will have to live with it. There is also a special notation for l. We write it as $\sum_{i=1}^{\infty} a_i$. Again from one point of view, this is a bad notation as it makes it look like an infinite process of addition, but on the other hand it is pretty natural once you get used to it and it works well from the following perspective:

$$l = \sum_{i=1}^{\infty} a_i = \lim_{n\to\infty} \sum_{i=1}^{n} a_i = \lim_{n\to\infty} s_n.^{4}$$

Just to be absolutely clear that we understand what $\sum_{i=1}^{\infty} a_i$ is I'll remind you that (if it exists) it is the real number that has the property that given any $\epsilon > 0$ there exists a natural number n_0 such that whenever $n > n_0$ we have $\left| \sum_{i=1}^{n} a_i - \sum_{i=1}^{\infty} a_i \right| < \epsilon$.

When we consider this, we might conclude that 'sum of a series' is not a bad name as we can get arbitrarily close to the limit by adding a large enough number of terms – so adding more terms beyond the N that takes you to within ϵ of the limit isn't going to give you much more if ϵ is sufficiently small! Indeed if the sum of the series exists, it is common for even the most rigorous mathematicians to write

$$\sum_{i=1}^{\infty} a_i = a_1 + a_2 + a_3 + \cdots \tag{6.3.5}$$

as though we really do have an infinite process of addition going on! There's no harm in doing this as long as you appreciate that (6.3.5) is nothing more than a suggestive notation. The truth is in the ϵs and Ns of the limit concept.

If (s_n) diverges to $+\infty$ we sometimes use the notation $\sum_{i=1}^{\infty} a_i = \infty$. Similarly we write $\sum_{i=1}^{\infty} a_i = -\infty$ when (s_n) diverges to $-\infty$. Also (and I hope this terminology won't confuse anyone), if I write that $\sum_{i=1}^{n} a_i$ (or even $\sum_{i=1}^{\infty} a_i$) converges (or diverges) this is sometimes just a convenient shorthand for the convergence (or divergence) of (s_n) where $s_n = \sum_{i=1}^{n} a_i$. So when we talk of *convergent* or *(divergent) series.* we really mean the convergence (or divergence) of the associated sequence of partial sums.

---

4 Just to confuse you, many textbooks write $\sum_{i=1}^{\infty} a_i$ as $\sum_{n=1}^{\infty} a_n$ which is perfectly OK as i and n are dummy indices, but which might be a little strange at first because of the role we have given to n. For this reason I'm sticking to i, at least in this chapter where the ideas are new to you!

Now what about some examples? To make things simpler at the beginning, for the next two sections we'll focus on *nonnegative series*, i.e. those for which $a_i \geq 0$ for each i – so for example (6.1.1) is included, but not (6.1.2).

6.4 Nonnegative Series

OK – so far all we've really done is to give a definition and introduce some new notation. Now it's time to get down to the serious business of real mathematics. I'll remind you that for the next few sections we're going to focus on sequences (a_n) where each $a_n \geq 0$. We consider the partial sums (s_n). Our first observation is that this is a monotonic increasing sequence as

$$s_{n+1} - s_n = \sum_{i=1}^{n+1} a_i - \sum_{i=1}^{n} a_i = \sum_{i=1}^{n} a_i + a_{n+1} - \sum_{i=1}^{n} a_i = a_{n+1} \geq 0,$$

and so $s_{n+1} \geq s_n$ for all n. It then follows from Corollary 5.2.1 that (s_n) either converges (to its supremum) or diverges to $+\infty$. Now we'll study some examples.

We've seen already that $\sum_{i=1}^{n} i = \frac{1}{2}n(n+1)$ and this clearly diverges to $+\infty$. If you look at higher powers of i then their partial sums are even larger and so we should expect divergence again, e.g. $\sum_{i=1}^{n} i^2 = 1 + 2^2 + 3^2 + \cdots + n^2 > n^2 > n$, and since (n) diverges, then so does $\sum_{i=1}^{n} i^2$. Indeed this is a consequence of Theorem 4.1.3. A similar argument applies to $\sum_{i=1}^{n} i^m$ where $m > 1$ is any real number. What about $m = 0$? Well here we have $\sum_{i=1}^{n} i^0 = \sum_{i=1}^{n} 1 = n$ which again diverges. Finally if $0 < m < 1$, we have $i^m \geq 1$ and so $\sum_{i=1}^{n} i^m \geq \sum_{i=1}^{n} 1 = n$ which also diverges. So we can conclude that $\sum_{i=1}^{n} i^m$ diverges to $+\infty$ for all $m \geq 0$. What happens when $m < 0$? Let's start by looking at the case $m = -1$. This is the famous *harmonic series* which is obtained by summing the terms of the harmonic sequence that we discussed in Section 4.1:[5]

$$\sum_{i=1}^{n} \frac{1}{i} = 1 + \frac{1}{2} + \frac{1}{3} + \frac{1}{4} + \cdots + \frac{1}{n}.$$

[5] For the relationship with the notion of harmonic in music see e.g. http://en.wikipedia.org/wiki/Harmonic_series_(music)

A reasonable conjecture might be that this series converges as $\lim_{n\to\infty} \frac{1}{n} = 0$ so as n gets very large the difference between s_n and s_{n+1} is getting smaller and smaller. Indeed if we calculate the first few terms we find that $s_1 = 1, s_2 = \frac{3}{2}, s_3 = \frac{11}{6}$, $s_4 = \frac{25}{12}, s_5 = \frac{137}{60}, \ldots$, so we might well believe that the sum of the series is a number between 2 and 3. But (as we should know by now), looking at the first five terms (or even the first billion) may not be a helpful guide to understanding convergence. In fact $\sum_{i=1}^{n} \frac{1}{i}$ diverges. To see why this is so, we'll employ a clever argument that collects terms together in powers of 2. We look at,[6]

$$\sum_{i=1}^{2^n} \frac{1}{i} = 1 + \frac{1}{2} + \left(\frac{1}{3} + \frac{1}{4}\right)$$

$$+ \left(\frac{1}{5} + \frac{1}{6} + \frac{1}{7} + \frac{1}{8}\right)$$

$$+ \left(\frac{1}{9} + \frac{1}{10} + \frac{1}{11} + \frac{1}{12} + \frac{1}{13} + \frac{1}{14} + \frac{1}{15} + \frac{1}{16}\right)$$

$$+ \cdots\cdots$$

$$+ \left(\frac{1}{2^{n-1}+1} + \frac{1}{2^{n-1}+2} + \cdots + \frac{1}{2^n}\right) \ldots (*)$$

Now observe that

$$\frac{1}{3} + \frac{1}{4} > \frac{1}{4} + \frac{1}{4} = \frac{1}{2},$$

$$\frac{1}{5} + \frac{1}{6} + \frac{1}{7} + \frac{1}{8} > \frac{1}{8} + \frac{1}{8} + \frac{1}{8} + \frac{1}{8} = \frac{1}{2},$$

$$\frac{1}{9} + \frac{1}{10} + \frac{1}{11} + \frac{1}{12} + \frac{1}{13} + \frac{1}{14} + \frac{1}{15} + \frac{1}{16} > 8.\frac{1}{16} = \frac{1}{2},$$

and we continue this argument until we get to

$$\frac{1}{2^{n-1}+1} + \frac{1}{2^{n-1}+2} + \cdots + \frac{1}{2^n} = \frac{1}{2^{n-1}+1} + \frac{1}{2^{n-1}+2} + \cdots + \frac{1}{2^{n-1}+2^{n-1}}.$$

There are 2^{n-1} terms in this general sum and each term is greater than $\frac{1}{2^n}$ so

$$\frac{1}{2^{n-1}+1} + \frac{1}{2^{n-1}+2} + \cdots + \frac{1}{2^n} > 2^{n-1}.\frac{1}{2^n} = \frac{1}{2}.$$

If we count the number of brackets on the right-hand side of (*) and also include the terms 1 and $\frac{1}{2}$, we find that we have $(n+1)$ terms altogether and we have seen that n of these terms exceed $\frac{1}{2}$.

[6] This argument is due to Nicole Oresme (1323?–1382), a Parisian thinker who eventually became Bishop of Lisieux – see e.g. http://en.wikipedia.org/wiki/Nicole_Oresme

We conclude that $\sum_{i=1}^{2^n} \frac{1}{i} > 1 + n.\frac{1}{2} = 1 + \frac{n}{2}$ and we can see from this that the series diverges. Indeed given any $K > 0$ we can find n_0 such that $1 + \frac{n}{2} > K$ for all $n > n_0$ (just take n_0 to be the smallest natural number larger than $2(K-1)$) and then $\sum_{i=1}^{n} \frac{1}{i} > K$ for all $n > 2^{n_0}$.

At this stage you might be getting the feeling that all series are divergent. You can rest assured that that is far from the case. There are plenty of convergent series around as we'll soon see. The next series on our list that we should consider is $\sum_{i=1}^{n} \frac{1}{i^2}$, but we need a few more tools before we can investigate that one. In fact we'll need to know about the related series $\sum_{i=1}^{n} \frac{1}{i(i+1)}$ and this will furnish our first example of a convergent series.

Example 6.1: $\displaystyle\sum_{i=1}^{n} \frac{1}{i(i+1)}$

To show this series converges we'll first find a neat formula for the nth partial sum. To begin with, you should check by cross-multiplication that

$$\frac{1}{i(i+1)} = \frac{1}{i} - \frac{1}{i+1}.$$

Next we write $\sum_{i=1}^{n} \frac{1}{i(i+1)}$ as a 'telescopic sum':[7]

$$\sum_{i=1}^{n} \frac{1}{i(i+1)} = \sum_{i=1}^{n} \left(\frac{1}{i} - \frac{1}{i+1}\right)$$

$$= \left(1 - \frac{1}{2}\right) + \left(\frac{1}{2} - \frac{1}{3}\right) + \left(\frac{1}{3} - \frac{1}{4}\right) + \cdots$$

$$+ \left(\frac{1}{n-2} - \frac{1}{n-1}\right) + \left(\frac{1}{n-1} - \frac{1}{n}\right) + \left(\frac{1}{n} - \frac{1}{n+1}\right)$$

$$= 1 - \frac{1}{n+1},$$

after cancellation. So we conclude that

$$\sum_{i=1}^{n} \frac{1}{i(i+1)} = 1 - \frac{1}{n+1}.$$

[7] So called because the series can be compressed into a simple form by cancellation. The analogy is with the collapse of an old-fashioned telescope.

However, we know that the sequence whose nth term is $\frac{1}{n+1}$ converges to 0 and so we find that

$$\sum_{i=1}^{\infty} \frac{1}{i(i+1)} = \lim_{n \to \infty} \sum_{i=1}^{n} \frac{1}{i(i+1)} = 1 - \lim_{n \to \infty} \frac{1}{n+1} = 1.$$

This is our first successful encounter with a convergent series so we should allow ourselves a quick pause for appreciation. In this case we also have an example where the sum of the series is explicitly known. This is in fact quite rare. In most cases where we can show that a series converges we will not know the limit explicitly.

At this stage it is worth thinking a little bit about the relationship between the sequences (a_n) and (s_n) from the point of view of convergence. The last two examples we've considered are $\sum_{i=1}^{n} \frac{1}{i}$ and $\sum_{i=1}^{n} \frac{1}{i(i+1)}$. In the first of these we have $a_n = \frac{1}{n}$ and we know that $\lim_{n \to \infty} \frac{1}{n} = 0$, but we've shown that (s_n) diverges. In the second example, $a_n = \frac{1}{n(n+1)}$ and again we have $\lim_{n \to \infty} \frac{1}{n(n+1)} = 0$, but in this case (s_n) converges as we've just seen. It's time for a theorem:

Theorem 6.4.1. If $\sum_{i=1}^{n} a_i$ converges then so does the sequence (a_n) and $\lim_{n \to \infty} a_n = 0$.

Proof. Suppose that $\lim_{n \to \infty} s_n = l$ then we also have $\lim_{n \to \infty} s_{n-1} = l$ (think about it). Now since

$$a_n = s_n - s_{n-1},$$

for all $n \geq 2$, we can use the algebra of limits, firstly to deduce that (a_n) converges and secondly to find that

$$\lim_{n \to \infty} a_n = \lim_{n \to \infty} s_n - \lim_{n \to \infty} s_{n-1}$$

$$= l - l = 0,$$

and we are done. □

It's important to appreciate what Theorem 6.4.1 is really telling us. It says that *if* (s_n) converges *then* it follows that (a_n) converges to zero. It should not be confused with the converse statement: '*if* (a_n) converges to zero *then* it follows that (s_n) converges' which is false – and the case $a_n = \frac{1}{n}$ provides a counter-example. On the other hand, one of the most useful applications of Theorem 6.4.1 is to prove that a series diverges, for if we use the fact that a statement is logically equivalent to its contrapositive,[8] we see that Theorem 6.4.1 also tells us that if (a_n)

---

[8] The *contrapositive* of 'If P then Q,' is 'If not Q then not P'.

does not converge to zero then $\sum_{i=1}^{n} a_i$ diverges, e.g. we see immediately that $\sum_{i=1}^{n} \frac{i}{i+100}$

diverges since $\lim_{n\to\infty} \frac{n}{n+100} = \lim_{n\to\infty} \frac{1}{1+\frac{100}{n}} = 1$ by algebra of limits.

By the way, we said that we'd only consider nonnegative series in this section, but you can check that Theorem 6.4.1 holds without this constraint.

6.5 The Comparison Test

The theory of series is full of *tests for convergence* which are various tricks that have been developed over the years for showing that a series is convergent or divergent. We'll meet a small number of these in this chapter. In fact we can't proceed further in our quest to show convergence of $\sum_{i=1}^{n} \frac{1}{i^2}$ without being able to use the comparison test, and we'll present this as a theorem. The proof of (1) is particularly sweet as it makes use of old friends from earlier chapters.

Theorem 6.5.1 (The Comparison Test). Suppose that (a_n) and (b_n) are sequences with $0 \le a_n \le b_n$ for all n. Then

1. if $\sum_{i=1}^{n} b_i$ converges then so does $\sum_{i=1}^{n} a_i$,

2. if $\sum_{i=1}^{n} a_i$ diverges then so does $\sum_{i=1}^{n} b_i$.

Proof. Throughout this proof we'll write $s_n = \sum_{i=1}^{n} a_i$ (as usual) and $t_n = \sum_{i=1}^{n} b_i$.

1. We are given that the sequence (t_n) converges and so it is bounded by Theorem 4.2.1. In particular it is bounded above and since each t_n is positive, it follows that there exists $K > 0$ such that $t_n = |t_n| \le K$ for all n. Now since each $a_i \le b_i$ we have for all n that

$$s_n = \sum_{i=1}^{n} a_i \le \sum_{i=1}^{n} b_i = t_n \le K,$$

and so the sequence (s_n) is bounded above. We have already pointed out that (s_n) is monotonic increasing and so we can apply Theorem 5.2.1 (1) to conclude that (s_n) converges as required.

2. This is really just a special case of Theorem 4.1.3, but since I didn't prove that result I will do so for this one. The sequence (s_n) diverges so given any $L > 0$ there exists a natural number n_0 such that $s_n > L$ whenever $n > n_0$. But since each $b_i \geq a_i$ we can argue as in (1) to deduce that $t_n \geq s_n > L$ for such n and hence (t_n) diverges. □

We'll now (as promised) apply the comparison test to show that $\sum_{i=1}^{n} \frac{1}{i^2}$ converges.

Example 6.2: $\sum_{i=1}^{n} \frac{1}{i^2}$

We know from Example 6.1 that $\sum_{i=1}^{n} \frac{1}{i(i+1)}$ converges and so by the algebra of limits, so does $2\sum_{i=1}^{n} \frac{1}{i(i+1)} = \sum_{i=1}^{n} \frac{2}{i(i+1)}$.

Now for all natural numbers i,

$$\frac{2}{i(i+1)} - \frac{1}{i^2} = \frac{1}{i}\left[\frac{2}{i+1} - \frac{1}{i}\right]$$

$$= \frac{1}{i}\left[\frac{2i - (i+1)}{i(i+1)}\right]$$

$$= \frac{i-1}{i^2(i+1)} \geq 0.$$

So if we take $a_i = \frac{1}{i^2}$ and $b_i = \frac{2}{i(i+1)}$, we have $0 \leq a_i \leq b_i$ and so by Theorem 6.5.1 (1), $\sum_{i=1}^{n} \frac{1}{i^2}$ converges.

We've now shown that $\sum_{i=1}^{n} \frac{1}{i^2}$ converges. But is it possible to find an exact value for the sum of this series? The problem was first posed by Jacob Bernoulli (1654–1705).[9] As Bernoulli was living in Basle, Switzerland at the time the problem of finding a real number k such that $\sum_{i=1}^{\infty} \frac{1}{i^2} = k$ became known as the 'Basle problem'. The problem was solved by one of the greatest mathematicians of all time,

---

[9] He was one of three brothers who all made important mathematical contributions. Furthermore, the sons and grandsons of this remarkable trio produced another five mathematicians – see e.g. http://en.wikipedia.org/wiki/Bernoulli_family. Be aware that Jacob is sometimes called by his Anglicised name James and should not be confused with his younger brother Johann (also called John).

Leonhard Euler[10] (1706–90) in 1735. He showed that

$$\sum_{i=1}^{\infty} \frac{1}{i^2} = \frac{\pi^2}{6},$$

so that $k = \frac{\pi^2}{6}$ which is 1.6449 to four decimal places. This connection between the sums of inverses of squares of natural numbers and π – the universal constant which is the ratio of the circumference of any circle to its diameter – is really beautiful and may appear a little mysterious. To give Euler's original proof goes beyond the scope of this book.[11] If you take a university level course that teaches you the idea of a Fourier series then there is a very nice succinct proof which uses that concept in a delightful way.

Now that we have the comparison test up our sleeves, then we can make much more progress in our goal to fully understand when $\sum_{i=1}^{n} \frac{1}{i^r}$ converges for various values of r.

Example 6.3: $\displaystyle\sum_{i=1}^{n} \frac{1}{i^r}$ for $r > 2$.

All the series of this type converge. To see this it's enough to notice that if i is any natural number then whenever $r > 2$

$$i^r \geq i^2.$$

Now by (L5) we have

$$\frac{1}{i^r} \leq \frac{1}{i^2},$$

and we can immediately apply the comparison test (Theorem 6.5.1 (1)) to deduce that $\sum_{i=1}^{n} \frac{1}{i^r}$ converges for $r > 2$.

Example 6.4: $\displaystyle\sum_{i=1}^{n} \frac{1}{i^r}$ for $0 < r < 1$.

In this case we have $i^r < i$ for all natural numbers i and so again by (L5), $\frac{1}{i} < \frac{1}{i^r}$. Now apply the comparison test in the form Theorem 6.5.1 (2) to deduce that $\sum_{i=1}^{n} \frac{1}{i^r}$ diverges for $0 < r < 1$.

---

[10] See http://en.wikipedia.org/wiki/Leonhard_Euler
[11] The best account of this that I've come across is in Jeffrey Stopple's superb textbook *A Primer of Analytic Number Theory*, Cambridge University Press (2003). See the Further Reading section for more about this book.

I've already told you how Euler found an exact formula for $\sum_{i=1}^{\infty} \frac{1}{i^2}$ which featured π^2. He also discovered exact formulae for $\sum_{i=1}^{\infty} \frac{1}{i^r}$ where r is an *even number* and these are all expressed in terms of π^r. I won't write down the exact formulae here as they are more complicated then the case $r = 2$. We'll postpone that to the next chapter as there is a fascinating connection with the number e which we're going to study there.[12] Remarkably, very little is still known about $\sum_{i=1}^{\infty} \frac{1}{i^r}$ in the case where r is an odd number ($r \geq 3$). We had to wait until 1978 for Roger Apéry (1916–94) to prove that $\sum_{i=1}^{\infty} \frac{1}{i^3}$ is an irrational number!

To complete the story of $\sum_{i=1}^{n} n^r$ where r is any real number, we should look at $\sum_{i=1}^{n} \frac{1}{i^r}$ for $1 < r < 2$. We'll need a new technique before we do that and this is the theme of the next section.

Before we do that, let's look at one more interesting series. We've shown that $\sum_{i=1}^{\infty} \frac{1}{i}$ diverges. Now we'll consider the sum of all reciprocals of the *square-free integers*:

$$1, 2, 3, 5, 6, 7, 10, 11, 13, 14, \ldots,$$

i.e. those natural numbers that can never have a perfect square as a factor.

It seems that this is a 'smaller sum' so it may be possible that it converges. We'll denote the generic square-free integer as i_{sf} where the suffix *sf* stands for 'square-free' and consider $\sum_{i_{sf}=1}^{\infty} \frac{1}{i_{sf}}$.[13] Remember that in Chapter 1, we showed that any natural number n can be written as $n = i_{sf} m^2$ where $m \leq n$ is a natural number.

Theorem 6.5.2. $\displaystyle\sum_{i_{sf}=1}^{\infty} \frac{1}{i_{sf}}$ diverges.

Proof. The result more or less follows from the inequality

$$\sum_{i=1}^{n} \frac{1}{i} \leq \left(\sum_{i_{sf}=1}^{n} \frac{1}{i_{sf}} \right) \left(\sum_{m=1}^{n} \frac{1}{m^2} \right).$$

---

[12] If you can't wait, try looking in Section 6.2 of Stopple's book which is cited in footnote 11.

[13] It may be that a better notation for this is $\sum_{i_{sf} < \infty} \frac{1}{i_{sf}}$, which we define to mean precisely $\lim_{n \to \infty} \sum_{i_{sf} < n} \frac{1}{i_{sf}}$.

To see that this holds you need only use the fact that for any $1 \leq i \leq n$, $\frac{1}{i} = \frac{1}{i_{sf}} \frac{1}{m^2}$ where $m \leq n$ and $i_{sf} \leq n$, so $\frac{1}{i}$ certainly appears in the product of the sums on the right-hand side – and that is enough. We've seen that $\sum_{m=1}^{\infty} \frac{1}{m^2}$ converges (to a positive number which is the supremum of the sequence of partial sums) and so we can make the right-hand side of the last inequality larger to obtain

$$\sum_{i=1}^{n} \frac{1}{i} \leq \left(\sum_{i_{sf}=1}^{n} \frac{1}{i_{sf}} \right) \left(\sum_{m=1}^{\infty} \frac{1}{m^2} \right) = k \sum_{i_{sf}=1}^{n} \frac{1}{i_{sf}},$$

where $k = \sum_{m=1}^{\infty} \frac{1}{m^2}$ (which we pointed out earlier is in fact equal to $\frac{\pi^2}{6}$).

From this and Theorem 4.1.3 we see that the divergence of $\sum_{i_{sf}=1}^{\infty} \frac{1}{i_{sf}}$ follows from that of $\sum_{i=1}^{\infty} \frac{1}{i}$. $\square$

Corollary 6.5.1. There are infinitely many square-free integers.

Proof. Suppose by way of contradiction that there was a largest square-free integer I_{sf}. Then

$$\sum_{i_{sf}=1}^{\infty} \frac{1}{i_{sf}} = 1 + \frac{1}{2} + \frac{1}{3} + \cdots + \frac{1}{I_{sf}},$$

and this is a finite sum of numbers. This contradicts the result of Theorem 6.5.2. $\square$

Since all prime numbers are square-free you may wonder whether the sum of all $\frac{1}{p}$ (where p is prime) converges or diverges. We'll come back to settle that question in Chapter 8.

6.6 Geometric Series

In this section we'll look at series such as

$$1 + \frac{1}{2} + \frac{1}{4} + \frac{1}{8} + \frac{1}{16} + \cdots, \tag{6.6.6}$$

$$6 + 18 + 54 + 162 + 486 + \cdots \tag{6.6.7}$$

Both of these are examples of *geometric series*, i.e. they are of the form $\sum_{i=0}^{\infty} ar^i = a + ar + ar^2 + ar^3 + \cdots$ The number a is (surprise, surprise) called the *first term* and r is called the *common ratio*.[14] So in (6.6.6) $a = 1$ and $r = \frac{1}{2}$. In (6.6.7), $a = 6$ and $r = 3$. You might have guessed that (6.6.7) diverges and speculated that (6.6.6) may well converge. To investigate this in the general case we'll first find a general formula for the nth partial sum $s_n = \sum_{i=0}^{n} ar^n$.[15] This finite series is sometimes called a *geometric progression*. To find the general formula we first note that if $r = 1$ then

$$s_n = a + a + \cdots + a = (n+1)a. \tag{6.6.8}$$

Now assume that $r \neq 1$ to find that

$$s_n = a + ar + ar^2 + \cdots + ar^n. \tag{6.6.9}$$

Then multiply both sides of (6.6.9) by r to obtain

$$rs_n = ar + ar^2 + ar^3 + \cdots + ar^n + ar^{n+1}. \tag{6.6.10}$$

Notice that all the terms on the right-hand sides of (6.6.9) and (6.6.10) are common to both equations, except the first term a in (6.6.9) and the last term ar^{n+1} in (6.6.10). If we subtract (6.6.10) from (6.6.9) we obtain

$$(1-r)s_n = a - ar^{n+1},$$

and since $r \neq 1$, it is legitimate to divide both sides by $1-r$ to get the formula we are seeking:

$$s_n = \frac{a(1 - r^{n+1})}{1-r}. \tag{6.6.11}$$

For example, you can use this to quickly calculate the sum of the first 10 terms in (6.6.6) by spotting that $a = 1$ and $r = \frac{1}{2}$ in this case. So we want

$$s_9 = \frac{1.\left(1 - \left(\frac{1}{2}\right)^{10}\right)}{1 - \frac{1}{2}} = 2 - \frac{1}{2^9} = 2 - \frac{1}{512} = \frac{1023}{512}.$$

You can use a similar argument to deduce that $s_n = 2 - \frac{1}{2^n}$ and this sequence clearly converges to 2.

More generally if $|r| < 1$ (so $-1 < r < 1$) then, by Example 4.2, we have $\lim_{n \to \infty} r^n = 0$. Hence if we apply the algebra of limits in (6.6.11) we see that the geometric series converges and

$$\sum_{i=0}^{\infty} ar^n = \frac{a}{1-r}. \tag{6.6.12}$$

---

[14] To understand why the word 'geometric' is used here go to the section "Relationship to geometry and Euclid's work" in http://en.wikipedia.org/wiki/Geometric_progression

[15] Note that we start at $i = 0$ here instead of $i = 1$, but this introduces no new difficulties in finding limits. If you want to you can even systematically replace $\sum_{i=0}^{n} ar^i$ with $\sum_{i=1}^{n+1} ar^{i-1}$. However do bear in mind that s_n is now the sum of the first $n+1$ terms.

In fact it is easy to check using (6.6.8) and (6.6.11) that the geometric series converges for no other value of r. Indeed if $r \geq 1$ it diverges to $+\infty$, if $r = -1$ it oscillates finitely between 0 and a and if $r < -1$ it oscillates infinitely.

Now we are going to use the geometric series as a tool in proving that $\sum_{i=1}^{n} \frac{1}{i^r}$ converges for $1 < r < 2$. We'll use a similar trick to the one we employed to prove that $\sum_{i=1}^{n} \frac{1}{i}$ diverges. We'll write the natural numbers in the form

$$2^0, 2^0 + 1, 2^1, 2^1 + 1, 2^2, 2^2 + 1, 2^2 + 2, 2^3 - 1, 2^3, \ldots$$

Now consider all the terms written in this form that lie between 2^{i-1} and $2^i - 1$ (including these two 'end-points') where i is an arbitrary natural number. There are exactly 2^{i-1} such numbers (think about it – and look at the case $i = 3$ from the list above for inspiration if necessary). Now define

$$b_i = \frac{1}{(2^{i-1})^r} + \frac{1}{(2^{i-1} + 1)^r} + \frac{1}{(2^{i-1} + 2)^r} + \cdots + \frac{1}{(2^i - 1)^r}.$$

Since $\frac{1}{(2^{i-1})^r}$ is the largest number which appears on the right-hand side we get

$$b_i \leq 2^{i-1} \cdot \frac{1}{(2^{i-1})^r} = \left(\frac{1}{2^{i-1}}\right)^{r-1} = \left(\frac{1}{2^{r-1}}\right)^{i-1},$$

and so $\sum_{i=1}^{n} b_i \leq \sum_{i=1}^{n} \left(\frac{1}{2^{r-1}}\right)^{i-1}$.

But the series on the right-hand side is a geometric one having first term 1 and common ratio $\frac{1}{2^{r-1}} < 1$ since $r > 1$. So this geometric series converges and hence by the comparison test so does $\sum_{i=1}^{n} b_i$. But this series is nothing but $\sum_{i=1}^{n} \frac{1}{i^r}$ rewritten in a clever way that involves powers of 2. And that's it.[16]

We have now completed the story of $\sum_{i=1}^{\infty} i^r$ where r is a real number. We have shown that the series diverges if $r \geq -1$ and converges if $r < -1$. The 'parameter' r plays a similar role here to the temperature of substances like water. Water is frozen solid for temperatures below 0 degrees celsius but at that temperature it melts to form a liquid and this is called a 'phase transition' in physics. If we keep on increasing the temperature then the water stays liquid until we get to 100 degrees celsius, when it changes to a gas and this is a second phase transition. By analogy we can regard the value $r = -1$ as indicating a phase transition between the regions where the series $\sum_{i=1}^{n} i^r$ diverges and converges (respectively). This analogy

---

[16] In fact this is an example of *regrouping* a series (see Section 6.10) and to be completely rigorous, Theorem 6.10.1 should precede the argument we've just given.

may not be as far-fetched as it seems as analysis plays an important role in the mathematical modelling of real phase transitions.

6.7 The Ratio Test

In the last section we saw the benefits of the comparison test for proving convergence of series. Although it is a wonderful thing, it is by no means the only tool that we need for playing the series game and indeed there are many important series such as $\sum_{i=1}^{\infty} \frac{1}{i!}$ where it doesn't help at all.[17] In this section we'll develop another useful test called the *ratio test*.[18] Before we prove this, it will be helpful to make some general remarks about infinite series.

Suppose that we have a finite series $\sum_{i=1}^{n} a_i$ where each $a_i \geq 0$. We can split the sum at any intermediate point

$$\sum_{i=1}^{n} a_i = \sum_{i=1}^{m} a_i + \sum_{i=m+1}^{n} a_i, \tag{6.7.13}$$

Can we do the same for infinite sums? Suppose that $\sum_{i=1}^{n} a_i$ converges to l. Now consider the sequence whose nth term is $\sum_{i=m+1}^{n} a_i$. This series also converges by the comparison test as we have (using the notation of Theorem 6.5.1) $b_i \leq a_i$ for each i where $b_i = 0$ if $1 \leq i \leq m$ and $b_i = a_i$ if $i \geq m+1$. It is then natural to define

$$\sum_{i=m+1}^{\infty} a_i = \lim_{n \to \infty} \sum_{i=m+1}^{n} a_i.$$

Applying the algebra of limits in (6.7.13) we find that

$$\sum_{i=m+1}^{\infty} a_i = \lim_{n \to \infty} \left(\sum_{i=1}^{n} a_i - \sum_{i=1}^{m} a_i \right)$$

$$= \sum_{i=1}^{\infty} a_i - \sum_{i=1}^{m} a_i,$$

---

[17] Remember that $i! = i(i-1)(i-2) \cdots 3.2.1$.

[18] Sometimes called *d'Alembert's ratio test* in honour of the French thinker Jean d'Alembert (1717–83) who was the first to publish it – see http://en.wikipedia.org/wiki/Jean_le_Rond_d'Alembert

and so

$$\sum_{i=1}^{\infty} a_i = \sum_{i=1}^{m} a_i + \sum_{i=m+1}^{\infty} a_i. \tag{6.7.14}$$

Notice from this that to show $\sum_{i=1}^{\infty} a_i$ converges it's enough to prove that $\sum_{i=m+1}^{\infty} a_i$ converges for any fixed m which can be as large as we like. This makes sense from an intuitive point of view as it's the 'tail' of the series, i.e. it's value beyond a certain point, that determines convergence.

We're now ready to describe the ratio test and as usual we'll state the test as a theorem. The proof of this gives us another opportunity to appreciate the value of geometric series.

Theorem 6.7.1 (The Ratio Test). Suppose that (a_n) is a sequence of positive real numbers for which $\lim_{n\to\infty} \frac{a_{n+1}}{a_n} = l$, then

- $\sum_{i=1}^{n} a_i$ converges if $l < 1$.
- $\sum_{i=1}^{n} a_i$ diverges if $l > 1$.
- If $l = 1$ the test is inconclusive and the series may converge or diverge.

Proof. First suppose that $l < 1$ and notice that we can then find an $\epsilon > 0$ such that

$$l + \epsilon < 1 \ldots \text{(i)},$$

indeed no matter how close to 1 the real number l may get, there is always some gap and choosing $\epsilon = \frac{1}{2}(1 - l)$, for example, will only fill half of that gap.

Now let's go back to the definition of the limit of a sequence. Since $\left(\frac{a_{n+1}}{a_n}\right)$ converges to l and given the value of ϵ that we've just chosen to satisfy (i), we know there exists a natural number n_0 such that if $n > n_0$ then $\left|\frac{a_{n+1}}{a_n} - l\right| < \epsilon$, i.e.

$$l - \epsilon < \frac{a_{n+1}}{a_n} < l + \epsilon \ldots \text{(ii)}$$

Now we can write each

$$a_n = \frac{a_n}{a_{n-1}} \frac{a_{n-1}}{a_{n-2}} \cdots \frac{a_{n_0+2}}{a_{n_0+1}} a_{n_0+1}.$$

Each ratio on the right-hand side satisfies (ii) and so we have

$$a_n < (l + \epsilon)^{n-n_0-1} a_{n_0+1},$$

since there are precisely $n - n_0 - 1$ ratios on the right-hand side. Now a_{n_0+1} is a fixed number and $\sum_{n=n_0+1}^{\infty} (l + \epsilon)^{n-n_0-1} a_{n_0+1}$ is a geometric series with first term a_{n_0+1} and common ratio $l + \epsilon$. This converges by (i) and so by the comparison test (bearing in mind (6.7.14)), $\sum_{n=1}^{\infty} a_n$ converges.

Now suppose that $l > 1$ and choose $\epsilon = l - 1$ in the left-hand side inequality of (ii). Then we can find m_0 such that if $m > m_0$ then $\frac{a_{m+1}}{a_m} > 1$, i.e. $a_{m+1} > a_m$. But then we cannot possibly have $\lim_{m \to \infty} a_m = 0$ and so $\sum_{n=1}^{\infty} a_n$ diverges by Theorem 6.4.1.[19]

To see that anything can happen when $l = 1$ consider how $\sum_{i=1}^{n} a_i$ behaves as $n \to \infty$ in the two cases $a_n = \frac{1}{n}$ and $a_n = \frac{1}{n^2}$. $\qquad \square$

Note that Theorem 6.7.1 assumes implicitly that we are dealing with a series for which $\lim_{n \to \infty} \frac{a_{n+1}}{a_n}$ exists. If it doesn't then we cannot apply the ratio test (at least not this form of it, see Exercise 6.11 for a more general version).

Example 6.5: $\displaystyle\sum_{i=1}^{\infty} \frac{x^i}{i!}$

We'll use the ratio test to examine the convergence of the series $\sum_{i=1}^{n} \frac{x^n}{i!}$ where x is an arbitrary positive number. This series will play an important role in the next chapter when we'll be learning about the irrational number that is denoted by e. The ratio test is easy to apply in this case, we have

$$a_n = \frac{x^n}{n!}, \quad a_{n+1} = \frac{x^{n+1}}{(n+1)!},$$

and so

$$\frac{a_{n+1}}{a_n} = \frac{x^{n+1}}{(n+1)!} \cdot \frac{n!}{x^n} = \frac{x \cdot n!}{(n+1)n!} = \frac{x}{n+1} \to 0 \text{ as } n \to \infty,$$

irrespective of the value of x. So $l = 0 < 1$ and we conclude that the series converges for all values of x. This proof has given us a lot. Not only do we know that $\sum_{i=1}^{n} \frac{1}{i!}$ converges (in fact, as we'll see in Chapter 7, the sum of the series is the special number e) but also that, e.g. $\sum_{i=1}^{n} \frac{102039^i}{i!}$ converges.

---

[19] Recall that I told you that this result would be helpful in proving divergence.

6.8 General Infinite Series

So far in this chapter we've only considered infinite series that have nonnegative terms. But what about more general series such as $\sum_{i=1}^{n}(-1)^i$ or $\sum_{i=1}^{n}\frac{x^i}{i!}$ where x is an arbitrary real number? Let's focus on the first of these. If we group the terms as

$$(-1+1)+(-1+1)+(-1+1)+\cdots$$

then it seems to be converging to zero but if we write it as

$$1+(-1+1)+(-1+1)+(-1+1)+\cdots$$

it looks like it converges to 1. But in fact this series diverges by Theorem 6.4.1. We should take this as a 'health warning' that dealing with negative numbers in infinite series might lead to headaches. We'll come back to *regrouping* terms in series later in this chapter. Now let's focus on the general picture. We are interested in the convergence (or otherwise) of a series $\sum_{i=1}^{n}a_i$ where (a_n) is an arbitrary sequence of real numbers, so we've dropped the constraint that these numbers are all nonnegative. It's a shame to lose all the knowledge we've gained in the early part of this chapter so let's introduce a link to that material. To each general series $\sum_{i=1}^{n}a_i$ we can associate the nonnegative series $\sum_{i=1}^{n}|a_i|$. Here's a key definition. The series $\sum_{i=1}^{n}a_i$ is said to be *absolutely convergent* if $\sum_{i=1}^{n}|a_i|$ converges. So, for example, the series $\sum_{i=1}^{n}(-1)^{i+1}\frac{1}{i^2}$ is absolutely convergent since $\sum_{i=1}^{n}|(-1)^{i+1}\frac{1}{i^2}|=\sum_{i=1}^{n}\frac{1}{i^2}$ converges. Now all we've done so far is make a definition. The next theorem tells us why this is useful.

Theorem 6.8.1. Any absolutely convergent series is convergent.

Proof. We want to show that the sequence whose nth term is $s_n=\sum_{i=1}^{n}a_i$ converges. Suppose that it is absolutely convergent. Then the sequence (t_n) converges where $t_n=\sum_{i=1}^{n}|a_i|$. Now each $|a_i|\geq a_i$ and $|a_i|\geq -a_i$ (recall Section 3.4) and so

$$0\leq a_i+|a_i|\leq 2|a_i|.$$

By algebra of limits, $2\sum_{i=1}^{n} |a_i|$ converges and hence by the comparison test so does $u_n = \sum_{i=1}^{n} (a_i + |a_i|)$. Then by algebra of limits again we have convergence of $s_n = u_n - t_n$ and our job is done. $\qquad\square$

By Theorem 6.8.1 we see immediately that $\sum_{i=1}^{n} (-1)^{i+1} \frac{1}{i^2}$ converges. Next we have an important example that picks up an earlier theme.

Example 6.6: $\displaystyle\sum_{i=1}^{\infty} \frac{x^i}{i!}$

We can now show that this series converges when x is an arbitrary real number. Indeed we've already shown this when $x \geq 0$. If $x < 0$ then

$$\sum_{i=1}^{n} \left| \frac{x^i}{i!} \right| = \sum_{i=1}^{n} \frac{|x^i|}{i!} = \sum_{i=1}^{n} \frac{|x|^i}{i!}$$

and since $|x| \geq 0$ this last series converges. So $\sum_{i=1}^{n} \frac{x^i}{i!}$ is absolutely convergent and hence is convergent by Theorem 6.8.1.

Both of the tests for convergence that we've met can be souped up into tests for general series. I'll state these but if you want proofs, you'll have to provide the details (see Exercise 6.12). Both cases are quite straightforward to deal with.

The Comparison Test – general case. If (a_n) is an arbitrary sequence of real numbers and (b_n) is a sequence of nonnegative numbers so that $|a_n| < b_n$ for all n then if $\sum_{i=1}^{n} b_n$ converges so does $\sum_{i=1}^{n} a_i$.

The Ratio Test – general case. If (a_n) is a sequence of real numbers for which $\lim_{n\to\infty} \left| \frac{a_{n+1}}{a_n} \right| = l$, then if $l < 1$, $\sum_{i=1}^{n} a_i$ converges, if $l > 1$, $\sum_{i=1}^{n} a_i$ diverges and if $l = 1$ then the test is inconclusive.

6.9 Conditional Convergence

We've seen in the last section that every absolutely convergent series is convergent. In this section we'll focus on the convergence of series that may not be absolutely convergent. For example, consider the series $\sum_{i=1}^{n} (-1)^{i+1} \frac{1}{i}$ which begins $1 - \frac{1}{2} + \frac{1}{3} - \frac{1}{4} + \frac{1}{5} - \cdots$ This series is certainly not absolutely convergent as we've already shown that the harmonic series $\sum_{i=1}^{n}\frac{1}{i}$ diverges. But the partial sums of this series will certainly be smaller than those of the harmonic series, so perhaps there is a chance that it will converge. Before we investigate further we'll need another definition. If (a_n) is a sequence of real numbers for which $\sum_{i=1}^{n}a_i$ converges but $\sum_{i=1}^{n}|a_i|$ diverges, then the series $\sum_{i=1}^{n}a_i$ is said to be *conditionally convergent*. We've not met any conditionally convergent series yet but the next theorem, which gives us another test for convergence, will give us the tool we need to find them. This convergence test is named after Gottfried Leibniz (1646–1716),[20] who was a renaissance man, par excellence! In his well-known book *Men of Mathematics*[21] that gives a series of short biographies of famous (male) mathematicians, E.T. Bell writes, 'Mathematics was one of the many fields in which Leibniz showed conspicuous genius: law, religion, statecraft, history, logic, metaphysics and speculative philosophy all owe to him contributions, any one of which would have secured his fame and preserved his memory'.

Theorem 6.9.1 (Leibniz' Test). Let (a_n) be a sequence of nonnegative numbers that is

(a) monotonic decreasing, (b) convergent to zero.

In this case the series $\sum_{i=1}^{n}(-1)^{i+1}a_i$ converges.

Proof. Let $s_n = \sum_{i=1}^{n}(-1)^{i+1}a_i$. Let n be even so that $n = 2m$ for some m. Then

$$s_{2m} = (a_1 - a_2) + (a_3 - a_4) + \cdots + (a_{2m-1} - a_{2m}).$$

It follows that

$$s_{2m+2} = s_{2(m+1)} = s_{2m} + (a_{2m+1} - a_{2m+2}) \geq s_{2m},$$

since (a_n) is monotonic decreasing. This shows that the sequence (s_{2n}) is monotonic increasing. It is also bounded above since

$$s_{2m} = a_1 - (a_2 - a_3) - (a_4 - a_5) - \ldots - (a_{2m-2} - a_{2m-1}) - a_{2m}$$

$$\leq a_1.$$

Here we've used the fact that the sequence (a_n) is nonnegative so that $a_{2m} \geq 0$ and that it is also monotonic decreasing, so each bracket is nonnegative. We can now apply Theorem 5.2.1 (1) to conclude that (s_{2m}) converges and we define $s = \lim_{m\to\infty} s_{2m}$. We now know that even partial sums converge. What about the odd ones? Well by algebra of limits we have

$$\lim_{m\to\infty} s_{2m+1} = \lim_{m\to\infty} s_{2m} + \lim_{m\to\infty} a_{2m+1} = s + 0 = s.$$

We've shown that $\lim_{m\to\infty} s_{2m} = s$ and $\lim_{m\to\infty} s_{2m+1} = s$. To prove the theorem we need to show that the full sequence (s_n) converges. It seems feasible that if it does, then $\lim_{n\to\infty} s_n = s$ and this is what we'll now prove. It's about time we had an ϵ and an n_0 again, so let's fix $\epsilon > 0$. Then from what has been proved above, there exists n_0 such that if $m > n_0$ then $|s_{2m} - s| < \epsilon$ and there exists n_1 such that if $m > n_1$ then $|s_{2m+1} - s| < \epsilon$. Now let $n > \max(n_0, n_1)$. Then either n is even and so $n = 2m$ for some m, or n is odd in which case $n = 2m + 1$. In either case we have $|s_n - s| < \epsilon$ and the result is proved. $\qquad\square$

It is very easy to apply Leibniz' test to see immediately that $\sum_{i=1}^{n}(-1)^{i+1}\frac{1}{i}$ converges and this gives us a nice example of a conditionally convergent series. In fact this is an example of a series where the sum is known and it is $\log_e(2)$ (the logarithm to base e of 2 whose decimal expansion begins 0.6931471). The proof uses calculus which goes beyond the scope of this book but I've included a sketch below for those who know some integration (and as an incentive to learn about it for those who don't).

We start with the following binomial series expansion which is valid for $-1 < x < 1$:

$$(1 + x)^{-1} = \sum_{n=0}^{\infty}(-1)^n x^n$$

$$= 1 - x + x^2 - x^3 + x^4 - \cdots$$

Now integrate both sides (the interchange of integration with infinite summation on the right-hand side needs justification) to get

$$\log_e(1 + x) = x - \frac{x^2}{2} + \frac{x^3}{3} - \frac{x^4}{4} + \cdots$$

$$= \sum_{n=1}^{\infty}(-1)^{n+1}\frac{x^n}{n}.$$

Because of the constraint on x we can't just put $x = 1$ (tempting though this may be) but after some careful work, it turns out that you can take the *limit* as $x \to 1$ (from below) and this gives the required result.

101

6.10 Regrouping and Rearrangements

There are two ways in which we can mix up the terms in an infinite series – by regrouping and by rearrangement. In the first of these we add the series in a different way by bracketing the terms differently (I did this earlier for the series $\sum_{i=1}^{n}(-1)^i$) but we don't change the order in which terms appear. In the second, we mix up the order in which terms appear as much as we please. For example, consider the series $\sum_{i=1}^{n}\frac{1}{n^2} = 1 + \frac{1}{4} + \frac{1}{9} + \frac{1}{16} + \frac{1}{25} + \cdots$. An example of a regrouping of this series is

$$1 + \left(\frac{1}{4} + \frac{1}{9}\right) + \left(\frac{1}{16} + \frac{1}{25} + \frac{1}{36}\right) + \cdots = 1 + \frac{13}{36} + \frac{469}{3600} + \cdots$$

and here is a rearrangement of the same series:

$$1 + \frac{1}{9} + \frac{1}{25} + \frac{1}{36} + \frac{1}{4} + \frac{1}{49} + \frac{1}{36} + \cdots$$

If a series is convergent then regrouping can't do it much harm but rearrangements can wreak havoc as we will see. First let's look at regrouping:

Theorem 6.10.1. If a series converges to l then it continues to converge to the same limit after any regrouping.

Proof. Suppose the series $\sum_{i=1}^{n}a_i$ converges to l. We'll write a general regrouping of the series as follows:

$$
\begin{aligned}
b_1 &= & a_1 + a_2 + \cdots + a_{m_1} \\
b_2 &= & a_{m_1+1} + a_{m_2+2} + \cdots + a_{m_2} \\
&\vdots& \\
b_n &= & a_{m_{n-1}+1} + a_{m_{n-1}+2} + \cdots + a_{m_n}.
\end{aligned}
$$

Then $\sum_{i=1}^{n}b_i = \sum_{r=1}^{m_n}a_r$ and we have $m_n \geq n$. Since the original series converges we know that given $\epsilon > 0$ there exists n_0 such that if $m_n > n_0$ then $\left|\sum_{r=1}^{m_n}a_i - l\right| < \epsilon$.

Now if $n > n_0$ we must have $m_n > n_0$ and so $\left|\sum_{i=1}^{n}b_i - l\right| < \epsilon$ which gives the required convergence. □

102

The contrapositive of the statement of this theorem tells us that if a series converges to more than one limit after regrouping then it diverges and this gives us another proof that $\sum_{i=1}^{n}(-1)^n$ diverges, as we've already shown how to group it in such a way that it converges to 0 and to 1.

Rearrangements are more complicated. We won't prove anything about them here but will be content with just stating two results which are both originally due to the nineteenth-century German mathematician Bernhard Riemann (1826–66):[22]

- If $\sum_{i=1}^{n}a_i$ is absolutely convergent to l then any rearrangement of the series is also convergent to the same limit.

- If $\sum_{i=1}^{n}a_i$ is conditionally convergent then given any real number x, it is possible to find a rearrangement such that the rearranged series converges to x. In fact rearrangements can even be found for which the series diverges to $+\infty$ or $-\infty$.

The second result quoted here is quite mindboggling and the two results taken together illustrate that there is quite a significant difference in behaviour between absolutely convergent and conditionally convergent series. To see a concrete example of how to rearrange conditionally convergent series to converge to different values look at pp. 177–8 of D.Bressoud *A Radical Approach to Real Analysis* (second edition), Mathematical Association of America (2007).[23]

We'll close this section with a remark about divergent series. You may think that once a series has been shown to diverge then that's the end of the story. In fact it can sometimes make sense to assign a number to a divergent series and even refer to it as the 'sum' – where 'sum' is interpreted in a different way from the usual. For example consider a slight variation on our old friend (6.1.2) – the series $\sum_{i=1}^{n}(-1)^{i+1}$. The great Leonhard Euler noticed that this is a geometric series with first term 1 and common ratio -1. Even though the formula (6.6.12) is not valid in this context, Euler applied it and argued that the series 'converges' to $\frac{1}{2}$. Euler's reasoning was incorrect here but his intuition was sound. If you redefine summation of a series to mean, taking the limit of averages of partial sums rather than partial sums themselves, then this is precisely the answer that you get. If you want to explore divergent series further from this point of view then a good place to start is Exercise 6.19. See also http://en.wikipedia.org/wiki/Divergent_series.

---

[22] See http://en.wikipedia.org/wiki/Bernhard_Riemann
[23] This book is briefly discussed in the Further Reading section.

6.11 Real Numbers and Decimal Expansions

In this book we've adopted a working definition of a real number as one that has a decimal expansion. But what do we really mean by this? Since $a_0.a_1a_2a_3a_4\cdots = a_0 + 0.a_1a_2a_3a_4\cdots$ if the number $a_0.a_1a_2a_3a_4\cdots$ is positive and $a_0 - 0.a_1a_2a_3a_4\cdots$ if it is negative, we might as well just consider those real numbers that lie between 0 and 1 for the purposes of this discussion. Now

$$0.a_1a_2a_3a_4\cdots = \frac{a_1}{10} + \frac{a_2}{100} + \frac{a_3}{1000} + \frac{a_4}{10^4} + \cdots$$

so we can see that decimal expansions are really nothing but a convenient shorthand for convergent infinite series of the form $\sum_{n=1}^{\infty} \frac{a_n}{10^n}$. As we've already remarked, a rational number either has a finite decimal expansion such as $\frac{1}{2} = \frac{5}{10} + \frac{0}{100} + \frac{0}{1000} + \cdots$ or the a_ns are given by a periodic (or eventually periodic) pattern such as $\frac{1}{3} = \sum_{n=1}^{\infty} \frac{3}{10^n}$ so each $a_n = 3$ in this example.

By the way, it appears that we have privileged the number 10 in this story but (as discussed in section 2.1) that is just a matter of convenience and collective habit. We could just as easily work in base 2 for example and represent all numbers between 0 and 1 as binary decimals $\sum_{n=1}^{\infty} \frac{b_n}{2^n}$. e.g. $\frac{1}{2} = 0.1$ in this base and $\frac{1}{3} = 0.\dot{0}\dot{1} = \sum_{n=1}^{\infty} \frac{1}{4^n}$. This is of course a geometric series and you should check that it has the right limit. We stick to base 10 from now on because we're used to it (recall the discussion in Section 2.1).

An interesting phenomenon occurs with numbers whose decimal expansion is always 9 after a given point, so they look like $x = a_1a_2\cdots a_N 999\cdots = a_1a_2\cdots a_N\dot{9} = a_1a_2\cdots a_N + \sum_{r=N+1}^{\infty} \frac{9}{10^r}$.

Let's focus on $\sum_{r=N+1}^{\infty} \frac{9}{10^r}$. This is a geometric series whose first term is $\frac{9}{10^{N+1}}$ and common ratio is $\frac{1}{10}$. So it converges to

$$\frac{\frac{9}{10^{N+1}}}{1 - \frac{1}{10}} = \frac{\frac{9}{10^{N+1}}}{\frac{9}{10}} = \frac{1}{10^N}.$$

This means that $x = a_1a_2\cdots a_N + 0.00\cdots 01$ where the 1 is in the Nth place after the decimal point. So we can write $x = a_1a_2\cdots a_{N'}$ where $a_{N'} = a_N + 1$, e.g. $0.367\dot{9} = 0.368$ and $0.999999\cdots = 0.\dot{9} = 1$.

Generally two distinct decimal expansions that differ in only one place give rise to different numbers. The phenomenon that occurs with repeating nines is a very special one where the notation breaks down and appears to be giving you two different numbers that are in fact identical. From a common-sense point of view this may be quite obvious as e.g. $\frac{1}{3} = 0.\dot{3}$ and $1 = 3 \times \frac{1}{3} = 3 \times 0.\dot{3} = 0.\dot{9}$.

Just before we conclude this chapter we may ask whether every series $\sum_{n=1}^{\infty} \frac{a_n}{10^n}$ converges to give a legitimate decimal expansion. Here the a_ns are chosen from $0, 1, 2, \ldots, 9$. It's easy to establish convergence. Since each $a_n \leq 9$ we can use the comparison test as $\sum_{n=1}^{\infty} \frac{9}{10^n} = 1$. Thus we have a complete correspondence between numbers that lie between 0 and 1 and infinite series of the form $\sum_{n=1}^{\infty} \frac{a_n}{10^n}$.

6.12 Exercises for Chapter 6

1. Investigate the convergence of the following series and find the sum whenever this exists

    (a) $\sum_{n=0}^{\infty} \frac{100}{3^n}$, (b) $\sum_{n=1}^{\infty} n(n+1)$ (c) $\sum_{n=1}^{\infty} \frac{4}{(n+1)(n+2)}$

    (d) $\sum_{n=1}^{\infty} \frac{1}{n(n+2)}$, (e) $\sum_{n=0}^{\infty} \sin^n(\theta)$ where $0 \leq \theta \leq 2\pi$.

2. Use known results about sequences to give thorough proofs that if $\sum_{n=1}^{\infty} a_n$ and $\sum_{n=1}^{\infty} b_n$ both converge then,

    (a) $\sum_{n=1}^{\infty} (a_n + b_n)$ converges,

    (b) $\lambda \sum_{n=1}^{\infty} a_n$ converges for all real numbers λ.

    Formulate and prove similar results which pertain to divergence in the case where both series are properly divergent to $+\infty$ or to $-\infty$. Why doesn't (a) extend to the case where one of the series is properly divergent to $+\infty$ while the other is properly divergent to $-\infty$?

3. Show that if $\sum_{n=1}^{\infty} a_n$ converges then $\lim_{N \to \infty} \sum_{n=N}^{\infty} a_n = 0$.

4. (a) Use geometric series to write the recurring decimal $0.\dot{1}\dot{7}$ as a fraction in its lowest terms. [Hint: $0.\dot{1}\dot{7} = \frac{17}{100} + \frac{17}{10000} + \frac{17}{10^6} + \cdots$.]

(b) Suppose that c is a block of m whole numbers in a recurrent decimal expansion $0.\dot{c}$ (e.g. if $c = 1234$ then $m = 4$ and $0.\dot{c} = 0.\dot{1}23\dot{4}$). Deduce that $0.\dot{c} = \frac{c}{10^m-1}$.

5. (a) Show that each of the series whose nth term[24] is given below diverges

$$\text{(i) } (1 + \epsilon)^n \text{ where } \epsilon > 0 \qquad \text{(ii) } \frac{n+1}{n+2}$$

(b) Find all values of ϵ for which the series whose nth term is $(-1 + \epsilon)^n$

(i) converges,   (ii) diverges.

6. Use the comparison test to investigate the convergence of the series whose nth term is as follows:

$$\text{(a) } \frac{1}{1+n^2}, \quad \text{(b) } \frac{1 + \cos(n)}{2^n}, \quad \text{(c) } \frac{n}{n^2-1}, \quad \text{(d) } \frac{n}{n^3+1}, \quad \text{(e) } \frac{2 + \sin(n)}{n}.$$

7. Use the ratio test to investigate the convergence of the series whose nth term is as follows:

$$\text{(a) } \frac{(n!)^2}{(2n)!}, \quad \text{(b) } \frac{2^n}{n!}, \quad \text{(c) } \frac{n!}{n^n}, \quad \text{(d) } \frac{n^n}{n!}.$$

Note that the solutions to (c) and (d) require some knowledge of the number e, which is discussed in Chapter 7.

8. Use any appropriate technique to investigate the convergence of the series whose nth term is as follows:

$$\text{(a) } \frac{1}{n^n}, \quad \text{(b) } \sqrt{n+1} - \sqrt{n}, \quad \text{(c) } \frac{1}{\sqrt{n(n+1)}}, \quad \text{(d) } \frac{n!(n+4)!}{(2n)!}.$$

Once again for (a) it helps if you know about e (or alternatively do Problem 13 first).

9. Show that if $\sum_{n=1}^{\infty} a_n$ is a convergent series of nonnegative real numbers and (b_n) is a bounded sequence of nonnegative real numbers, then the series $\sum_{n=1}^{\infty} a_n b_n$ also converges.

10. Show that if $\sum_{n=1}^{\infty} a_n$ is a convergent series of positive real numbers then the series $\sum_{n+1}^{\infty} \sqrt{a_n a_{n+1}}$ is also convergent.

[24] Here and below by the 'nth term of the series' is meant a_n in $\sum_{n=1}^{\infty} a_n$.

11. Prove the following more powerful form of the ratio test which does not assume that $\lim_{n\to\infty}\frac{a_{n+1}}{a_n}$ exists:

    Let (a_n) be a sequence of positive numbers for which there exists $0 \le r < 1$ and a whole number n_0 such that for all $n > n_0$, $\frac{a_{n+1}}{a_n} \le r$ then $\sum_{n=1}^{\infty} a_n$ converges.

    If $\frac{a_{n+1}}{a_n} > 1$ for all $n \ge n_0$ then $\sum_{n=1}^{\infty} a_n$ diverges.

12. Give thorough proofs of the comparison test and the ratio test for series of the form $\sum_{n=1}^{\infty} a_n$ where (a_n) is an arbitrary real-valued sequence.

13. Although they are the best known, the comparison test and the ratio test are not the only tests for convergent series. Another well-known test is *Cauchy's root test*. This states that if (a_n) is a sequence with each $a_n > 0$ and if there exists $0 < r < 1$ with $\sqrt[n]{a_n} < r$ for all n then $\sum_{n=1}^{\infty} a_n$ converges, but if $\sqrt[n]{a_n} > 1$ for all n then $\sum_{n=1}^{\infty} a_n$ diverges. To prove this result:

    (a) Assume $\sqrt[n]{a_n} < r$ with $0 < r < 1$. Show that $a_n \le r^n$ for all n and hence use the comparison test with a geometric series to show $\sum_{n=1}^{\infty} a_n$ converges.

    (b) Suppose that $\sqrt[n]{a_n} > 1$ for all n. Deduce that $\lim_{n\to\infty} a_n = 0$ cannot hold and hence show that $\sum_{n=1}^{\infty} a_n$ diverges.

    (c) Deduce the stronger form of the root test whereby for convergence we only require that $\sqrt[n]{a_n} < r$ for all $n \ge n_0$ where n_0 is a given whole number and for divergence we ask that $\sqrt[n]{a_n} > 1$ for infinitely many n.

14. In Section 6.6 we showed that $\sum_{n=1}^{\infty} \frac{1}{n^r}$ converges whenever $r > 1$. You can use a similar argument to prove another test for convergence of series called *Cauchy's condensation test*: if (a_n) is a nonnegative monotonic decreasing sequence, then $\sum_{n=1}^{\infty} a_n$ converges if and only if $\sum_{n=1}^{\infty} 2^n a_{2^n}$ converges.

    Hint: First show that for each natural number k,

    $$a_{2^k} + a_{2^k+1} + \ldots a_{2^{k+1}-1} \le 2^k a_{2^k},$$

    $$a_{2^k+1} + a_{2^k+2} + \ldots a_{2^{k+1}} \ge 2^k a_{2^{k+1}}.$$

15. For each of the following series, decide whether it is (a) convergent, (b) absolutely convergent, giving your reasons in each case.

    (i) $\displaystyle\sum_{n=1}^{\infty} \frac{(-1)^{n+1}}{\sqrt{n}}$, (ii) $\displaystyle\sum_{n=1}^{\infty} \frac{(-1)^{n-1}}{3n^2 - 2n}$, (iii) $\displaystyle\sum_{n=1}^{\infty} \frac{\cos(n\pi)}{n^5}$.

16. Only one of the following statements is true. Present either a counter-example or a proof in each case.

   (a) If $\sum_{n=1}^{\infty} a_n^2$ converges then so does $\sum_{n=1}^{\infty} |a_n|$.

   (b) If $\sum_{n=1}^{\infty} |a_n|$ converges then so does $\sum_{n=1}^{\infty} a_n^2$.

17. Give an example of a sequence (a_n) for which $\sum_{n=1}^{\infty} a_n$ converges but $\sum_{n=1}^{\infty} a_n^2$ does not.

18. Recall Cauchy's inequality for sums from Exercise 3.9: If $a_1, a_2, \ldots, a_n$ and $b_1, b_2, \ldots, b_n$ are real numbers then

$$\left| \sum_{i=1}^{n} a_i b_i \right| \leq \left(\sum_{i=1}^{n} a_i^2 \right)^{\frac{1}{2}} \left(\sum_{i=1}^{n} b_i^2 \right)^{\frac{1}{2}}.$$

Now extend this inequality to series, so if (a_n) and (b_n) are sequences and we are given that both $\sum_{n=1}^{\infty} a_n^2$ and $\sum_{n=1}^{\infty} b_n^2$ converge, show that $\sum_{i=1}^{n} a_i b_i$ converges absolutely and that

$$\sum_{n=1}^{\infty} |a_n b_n| \leq \left(\sum_{n=1}^{\infty} a_n^2 \right)^{\frac{1}{2}} \left(\sum_{n=1}^{\infty} b_n^2 \right)^{\frac{1}{2}}.$$

19. Suppose that $\sum_{n=1}^{\infty} a_n$ diverges. It is sometimes useful to find the sum of an associated series that really converges. One example of a summation technique for associating a convergent series to a divergent one is *Cèsaro summation*. Recall the sequence of partial sums (s_n) where $s_n = \sum_{i=1}^{n} a_i$. We define the *Cèsaro average* of the partial sums to be $s_n' = \frac{1}{n}(s_1 + s_2 + \cdots + s_n)$ and if $\lim_{n \to \infty} s_n'$ exists then we say that the series $\sum_{n=1}^{\infty} a_n$ is *Cèsaro summable*.

   (a) Prove that if $\sum_{n=1}^{\infty} a_n$ converges in the usual sense then it is also Cèsaro summable and that both limits are the same.

   (b) Let $a_n = (-1)^{n+1}$. Show that $\sum_{n=1}^{\infty} a_n$ is Cèsaro summable and that the limit of the Cèsaro averages is $\frac{1}{2}$.

Part II
Exploring Limits

7

Wonderful Numbers - e, π and γ

Thus e became the first number to be defined by a limiting process.
e: The Story of a Number, E. Maior

In this chapter we'll look at three really interesting real numbers. Of course all numbers are interesting, but these three are of particular importance because they are so prevalent throughout science and mathematics.

7.1 The Number e

Suppose that you invest some money into a bank account at an interest rate r per annum. For example if the rate is 6% then $r = 0.06$. The amount you invest is called the *principal* and we denote it by the letter P. After a year has passed you have earned rP interest and so your initial investment of P has grown to $P(1 + r)$. From now on we'll simplify as much as possible and take $P = 1$ (pound, dollar, euro, yen – whatever you like). Now suppose that instead of interest being paid after a year it is paid at six-monthly intervals. Then after six months, your initial investment has grown to $\left(1 + \frac{r}{2}\right)$. Suppose this money is reinvested for the next six months. Then after a year you will have $\left(1 + \frac{r}{2}\right)^2$. By similar arguments you can see that if interest is paid monthly then after a year you have $\left(1 + \frac{r}{12}\right)^{12}$, if it is paid daily (and we are not in a leap year) then your investment is worth $\left(1 + \frac{r}{365}\right)^{365}$ at the end of the year and if it is paid every minute then your money grows to $\left(1 + \frac{r}{525600}\right)^{525600}$. Now the reader should calculate all of these numbers in the simplest case where we take $r = 1$. What happens if interest is paid every second – and every hundredth of second? You should by now have realised that you are calculating terms in the sequence whose nth term is $\left(1 + \frac{1}{n}\right)^n$, and you should

110

have accumulated some persuasive evidence that the sequence is converging to a number that is reasonably close to $2.718281828\ldots$ But how do we know the limit really exists? We'll demonstrate that now.

Theorem 7.1.1. The sequence whose nth term is $\left(1 + \frac{1}{n}\right)^n$ converges to a real number which we call e. We have $2 < e \leq 3$.

Proof. We'll do this in stages.

Stage 1: Our sequence is monotonic increasing
   To prove this we'll use an old friend – the Theorem of the Means (Theorem 3.5.1) which told us that if we have n positive real numbers $a_1, a_2, \ldots, a_n$ then

$$\sqrt[n]{a_1 a_2 \cdots a_n} \leq \frac{a_1 + a_2 + \cdots + a_n}{n}.$$

Now we'll apply this result with $a_1 = a_2 = \cdots = a_{n-1} = 1 + \frac{1}{n-1}$ and $a_n = 1$. Then the geometric mean is $\left(1 + \frac{1}{n-1}\right)^{\frac{n-1}{n}}$ and the arithmetic mean is

$$\frac{(n-1)\left(1 + \frac{1}{n-1}\right) + 1}{n} = \frac{n+1}{n} = 1 + \frac{1}{n}.$$

So the theorem of the means yields tells us that

$$\left(1 + \frac{1}{n-1}\right)^{\frac{n-1}{n}} \leq 1 + \frac{1}{n},$$

and raising both sides to the power n gives

$$\left(1 + \frac{1}{n-1}\right)^{n-1} \leq \left(1 + \frac{1}{n}\right)^n,$$

which is exactly what we wanted to prove.

Stage 2: Our sequence is bounded
   To prove this we first use the binomial theorem to expand

$$\left(1 + \frac{1}{n}\right)^n = 1 + n.\frac{1}{n} + \frac{n(n-1)}{2!}\frac{1}{n^2} + \frac{n(n-1)(n-2)}{3!}\frac{1}{n^3} + \cdots + \frac{1}{n^n}.$$

Using a little bit of algebra we can rewrite this to get

$$\left(1 + \frac{1}{n}\right)^n = 1 + \frac{1}{1!} + \left(1 - \frac{1}{n}\right)\frac{1}{2!} + \left(1 - \frac{1}{n}\right)\left(1 - \frac{2}{n}\right)\frac{1}{3!}$$

$$+ \cdots + \left(1 - \frac{1}{n}\right)\left(1 - \frac{2}{n}\right)\cdots\left(1 - \frac{n-1}{n}\right)\frac{1}{n!} \qquad (7.1.1)$$

111

If you're worried about where the last term came from, observe that when you simplify the product of brackets you obtain

$$\left(1 - \frac{1}{n}\right)\left(1 - \frac{2}{n}\right)\cdots\left(1 - \frac{n-1}{n}\right)\frac{1}{n!} = \frac{(n-1)!}{n^{n-1}}\frac{1}{n!} = \frac{1}{n^n}.$$

The formula (7.1.1) is intriguing and we'll return to it when the proof is complete. For now just notice that all of the terms taking the form $\left(1 - \frac{k}{n}\right)$ which appear in the right-hand side of (7.1.1) are between 0 and 1 and so we can deduce that

$$\left(1 + \frac{1}{n}\right)^n \leq 1 + \frac{1}{1!} + \frac{1}{2!} + \cdots + \frac{1}{n!}$$

$$\leq \sum_{i=0}^{\infty} \frac{1}{i!} \tag{7.1.2}$$

Now we proved that the infinite series appearing on the right-hand side of (7.1.2) converged, in Section 6.7. So its sum is an upper bound for our sequence. As the sequence is bounded above and monotonic increasing we can assert that it converges, by Theorem 5.2.1. Thus we can now legitimately define

$$e = \lim_{n \to \infty} \left(1 + \frac{1}{n}\right)^n. \tag{7.1.3}$$

Now we need to show that $2 < e \leq 3$. Before we do that, it's helpful to take a quick diversion.

Stage 3: A Useful Inequality

Let's start by looking at factorials again: $n! = n(n-1)(n-2)\ldots 3.2.1$. There are n numbers multiplied together on the right-hand side and all but one of them is greater than or equal to 2. That tells us that $n! \geq 2^{n-1}$ and so by (L5)

$$\frac{1}{n!} \leq \frac{1}{2^{n-1}}. \tag{7.1.4}$$

Stage 4: The Bounds

From the first line of (7.1.1), we see that 2 is a lower bound for the sequence $\left(1 + \frac{1}{n}\right)^n$. Since e is the supremum of this sequence it then follows that $e > 2$. For the upper bound we first use the inequality just before (7.1.2) and then apply (7.1.4) to show that

$$\left(1 + \frac{1}{n}\right)^n \leq 1 + 1 + \frac{1}{2!} + \cdots + \frac{1}{n!}$$

$$\leq 1 + 1 + \frac{1}{2} + \frac{1}{2^2} + \cdots + \frac{1}{2^{n-1}}.$$

Now the geometric series $1 + \frac{1}{2} + \frac{1}{2^2} + \cdots + \frac{1}{2^{n-1}}$ has first term 1 and common ratio $\frac{1}{2}$ and so it converges to 2. Hence we see that 3 is an upper bound for the sequence with nth term $\left(1 + \frac{1}{n}\right)^n$. But e is the supremum of this sequence and so $e \leq 3$ as promised. $\square$

We have now defined e as the limit of a *sequence*. The formula (7.1.1) which helped us do this is intriguing. It suggests that we might also be able to write e as the limit of a *series*:[1]

$$e = 1 + \frac{1}{1!} + \frac{1}{2!} + \frac{1}{3!} + \cdots = \sum_{n=0}^{\infty} \frac{1}{n!} \qquad (7.1.5)$$

But beware. You may be tempted to try to prove this very quickly by taking limits on both sides of (7.1.1); we need to proceed with caution as the right-hand side is quite a complicated expression.

Theorem 7.1.2. $e = \sum_{n=0}^{\infty} \frac{1}{n!}$.

Proof. We see from (7.1.2) that $\sum_{n=0}^{\infty} \frac{1}{n!}$ is an upper bound for the sequence whose nth term is $\left(1 + \frac{1}{n}\right)^n$. But we proved in Theorem 7.1.1 that e is the supremum of this sequence and so

$$e \leq \sum_{n=0}^{\infty} \frac{1}{n!}.$$

Now consider (7.1.1) again but this time only take the first m terms on the right-hand side where $m < n$. It's a good idea to give this quantity a name so define

$$c_n^{(m)} = 1 + \frac{1}{1!} + \left(1 - \frac{1}{n}\right)\frac{1}{2!} + \left(1 - \frac{1}{n}\right)\left(1 - \frac{2}{n}\right)\frac{1}{3!}$$

$$+ \cdots + \left(1 - \frac{1}{n}\right)\left(1 - \frac{2}{n}\right)\cdots\left(1 - \frac{m-1}{n}\right)\frac{1}{m!}.$$

Then we have,[2]

$$c_n^{(m)} \leq \left(1 + \frac{1}{n}\right)^n \leq e.$$

We fix m for now and consider the sequence $(c_n^{(m)}) = (c_1^{(m)}, c_2^{(m)}, c_3^{(m)}, \ldots)$. We have just shown that it is bounded above by e. It is also monotonic increasing, indeed this follows easily from the fact that $1 - \frac{k}{n}$ is monotonic increasing for each

---

[1] Note that in (7.1.5) we define $0! = 1$ and this is standard throughout mathematics.
[2] m is playing the role of a label here and should not be confused with a power. That's why I've put it in brackets.

$1 \leq k \leq m - 1$. Thus we see that the sequence converges to $1 + 1 + \frac{1}{2!} + \cdots + \frac{1}{m!}$ as $n \to \infty$ and since this limit is also the supremum we must have

$$1 + 1 + \frac{1}{2!} + \cdots + \frac{1}{m!} \leq e.$$

Now the argument we've just used holds for arbitrary m. So we see that e is an upper bound for the sequence whose mth term is $\sum_{n=0}^{m} \frac{1}{m!}$ and it follows that

$$\sum_{n=0}^{\infty} \frac{1}{n!} \leq e.$$

Now we've shown that $e \leq \sum_{n=0}^{\infty} \frac{1}{n!}$ and $\sum_{n=0}^{\infty} \frac{1}{n!} \leq e$ and can only conclude that

$$e = \sum_{n=0}^{\infty} \frac{1}{n!}. \qquad \square$$

In fact the results of the last two theorems can be generalised. Suppose that x is an arbitrary real number. Then it can be shown that

$$e^x = \lim_{n \to \infty} \left(1 + \frac{x}{n}\right)^n = \sum_{n=0}^{\infty} \frac{x^n}{n!}.$$

The surprising thing here is not so much that the two limits exist and are equal to each other but that they are equal to e^x. This tells us for example, that

$$\sum_{n=0}^{\infty} \frac{2^n}{n!} = \left(\sum_{n=0}^{\infty} \frac{1}{n!}\right)^2,$$

which is far from obvious when you first see it. To see that this is true we really need to verify that

$$e^{x+y} = e^x e^y,$$

for all real numbers x and y.

For a sketch (and the reader is invited to fill in the gaps) of this important fact we need to use the binomial theorem (see Appendix 1) to write

$$e^{x+y} = \sum_{n=0}^{\infty} \frac{(x+y)^n}{n!}$$

$$= \sum_{n=0}^{\infty} \frac{1}{n!} \sum_{k=0}^{n} \binom{n}{k} x^k y^{n-k}$$

$$= \sum_{n=0}^{\infty} \sum_{k=0}^{n} \frac{1}{k!(n-k)!} x^k y^{n-k}$$

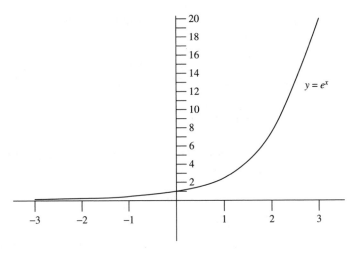

Figure 7.1. The exponential function.

$$= \left(\sum_{n=0}^{\infty} \frac{x^n}{n!} \right) \left(\sum_{m=0}^{\infty} \frac{y^m}{m!} \right)$$

$$= e^x e^y.$$

The curve that you obtain when you vary x in e^x is the graph of the *exponential function* (Figure 7.1).

This crops up time and time again in science, engineering, economics and many other areas where mathematics is applied to understand the real world. Indeed the expression for continuously compounded interest at time t that motivated the work of this chapter is given by Pe^{rt}. We could spend a lot of time on the function e^x but for this book it's enough to consider three things.

(i) If $x > 0$ then e^x is the limit (and hence the supremum) of a bounded monotonic increasing sequence. It follows that for all natural numbers n,

$$e^x \geq 1 + x + \frac{x^2}{2!} + \cdots + \frac{x^n}{n!}.$$

In particular, we'll often need the case where $n = 1$:

$$e^x \geq 1 + x. \qquad (7.1.6)$$

(ii) If $1 < x < y$ then $\frac{x^n}{n!} < \frac{y^n}{n!}$ for all n and it follows that $e^x < e^y$.

(iii) Sometimes we want to consider e^x when x is itself quite a complicated expression. In that case we will use the notation $\exp\{x\}$ in place of e^x so that we don't have to ask the reader to strain their eyes too much. Here 'exp' stands for *exponential*.

It's time for an interesting diversion. In Chapter 6, we looked at the finite series $\sum_{i=1}^{n} i = \frac{1}{2}n(n+1)$. If you have taken a first year undergraduate course in mathematics you might well have encountered the formulae (which are usually proved using the technique of mathematical induction (see Appendix 3)):

$$\sum_{i=1}^{n} i^2 = \frac{1}{6}n(n+1)(2n+1) \quad \text{and} \quad \sum_{i=1}^{n} i^3 = \frac{1}{4}n^2(n+1)^2.$$

There doesn't appear to be a pattern here but there is and it was found by Jacob Bernoulli. He showed that for any natural number m:

$$\sum_{i=1}^{n-1} i^m = \sum_{k=0}^{m} \frac{m!}{(m+1-k)!k!}B_k n^{m+1-k}.$$

You can sum up to n on the left-hand side if you like, but that means you have to replace n by $n+1$ on the right-hand side which makes it more complicated. The numbers $B_0, B_1, \ldots, B_m$ which appear in this formula are called *Bernoulli numbers* in honour of their discoverer. The remarkable thing is that they involve e^x, indeed the best way to define Bernoulli numbers is indirectly through the identity:

$$\frac{x}{e^x - 1} = \sum_{n=0}^{\infty} B_n \frac{x^n}{n!},$$

so by (7.1.2)

$$x = \left(\sum_{n=0}^{\infty} B_n \frac{x^n}{n!}\right)\left(\sum_{r=1}^{\infty} \frac{x^r}{r!}\right).$$

You can equate coefficients in this last identity to calculate the Bernoulli numbers, so if you compare coefficients of x on the left and the right you get $1 = B_0$. Similarly comparing coefficients of x^2 you find that $0 = \frac{1}{2} + B_1$ so $B_1 = -\frac{1}{2}$. Continuing in this way you can find $B_2 = \frac{1}{6}$, $B_3 = 0$, $B_4 = -\frac{1}{30}, \ldots,$[3] and hence calculate $\sum_{i=1}^{n-1} i^m$ for as high a value of m as you like. Incidentally it turns out that B_{2k+1} is always equal to zero and B_{2k+2} is always negative.

In Chapter 6, we mentioned Euler's remarkable result that $\sum_{n=1}^{\infty} \frac{1}{n^2} = \frac{\pi^2}{6} = B_2\pi^2$. He also proved the more general result

$$\sum_{n=1}^{\infty} \frac{1}{n^{2k}} = \frac{(-1)^{k-1}4^{k-\frac{1}{2}}}{(2k)!}B_{2k}\pi^{2k}.$$

[3] See http://en.wikipedia.org/wiki/Bernoulli_number

116

This should be enough to convince you that Bernoulli numbers are rather useful tools to have at your disposal.

Let's return to the number e itself. It must be either rational or irrational. The answer is in the following theorem:

Theorem 7.1.3. e is an irrational number.

Proof. We'll give a proof by contradiction, so we'll begin by assuming that e is a rational number and so it can be written as $\frac{p}{q}$ where p and q are natural numbers. Now for each $1 \leq k \leq n$, $\frac{n!}{k!}$ is a natural number, indeed you should check that $\frac{n!}{k!} = n(n-1)(n-2)\cdots(k+1)$. Also if $n \geq q$, we can certainly write $n! = mq$ for some natural number m. It follows that $n!e = n!\frac{p}{q} = mp$ is also a natural number. From now on we choose $n \geq q$. From what we've just written, we know that N is an integer where

$$N = n!e - \sum_{k=0}^{n} \frac{n!}{k!}.$$

In fact N is a natural number since

$$N = n!\left(e - \sum_{k=0}^{n} \frac{1}{k!}\right),$$

and we know that $e > \sum_{k=0}^{n} \frac{1}{k!}$. Now using Theorem 7.1.2 and (6.7.14) we get

$$N = n!\left(\sum_{k=0}^{\infty} \frac{1}{k!} - \sum_{k=0}^{n} \frac{1}{k!}\right)$$

$$= n! \sum_{k=n+1}^{\infty} \frac{1}{k!}$$

$$= \sum_{k=n+1}^{\infty} \frac{n!}{k!}$$

$$= \frac{1}{n+1} + \frac{1}{(n+1)(n+2)} + \cdots$$

$$= \sum_{k=1}^{\infty} \frac{1}{(n+1)(n+2)\cdots(n+k)}.$$

Now since $(n+1)(n+2)\cdots(n+k) > (n+1)^k$ for $k > 1$, we have

$$N < \sum_{k=1}^{\infty} \frac{1}{(n+1)^k}.$$

The series on the right-hand side is a geometric one with first term and common ratio both equal to $\frac{1}{n+1}$. So it converges to $\frac{\frac{1}{n+1}}{1-\frac{1}{n+1}} = \frac{1}{n}$, and so we conclude that $N < \frac{1}{n}$. This gives us our required contradiction. □

Now we know that e is irrational, is it perhaps one of those numbers that we have encountered previously such as the square root of a prime number? To answer this question we need a new definition.

Let x be a real number. We say that it is *algebraic* if there exists a natural number n and integers,[4] $a_0, a_1, \ldots, a_n$ such that

$$a_0 + a_1x + a_2x^2 + \cdots + a_nx^n = 0,$$

i.e. the number x is a solution of an equation with integer coefficients. What kinds of numbers are algebraic? Well to start with all rational numbers are, since if such a number can be written in the form $\frac{p}{q}$ then we simply take $n = 1$, $a_0 = -p$ and $a_1 = q$ to see that $-p + qx = 0$. If p is any prime number (or indeed any real number), then $\sqrt{p}$ is also algebraic as we see by taking $n = 2$, $a_0 = -p$, $a_1 = 0$ and $a_2 = 1$. Are there any irrational numbers that fail to be algebraic? The answer to this question is yes, but unfortunately the details required to prove it are too complex for a book of this nature. Numbers that fail to be algebraic are called *transcendental*. The first examples of such numbers were found by the French mathematician Joseph Liouville (1809–82) in 1844. In particular he proved that $\sum_{n=1}^{\infty} \frac{1}{10^{n!}} = 0.110001\ldots$ is transcendental. Another mathematician, Charles Hermite (1822–1901), proved that e is transcendental in 1873. His proof ran to more than thirty pages. Another well-known number that turns out to be transcendental is π and we will turn to this next.

7.2 The Number π

In this and the next section we'll depart from the main philosophy of this book which has been to avoid using calculus. For π and γ it's impossible to probe them with any degree of depth without needing to use more sophisticated tools and these sections should be treated by readers as an invitation to the feast rather than a course (not even a starter).

It has been known since antiquity that the ratio of the circumference to the diameter of any (idealised) circle is constant. By idealised here I mean that the circle cannot be drawn by any physical instrument as its boundary has zero thickness. This constant has been denoted by the Greek letter π since 1706

---

[4] In fact it's equivalent to take $a_0, a_1, \ldots, a_n$ to be rational. Why?

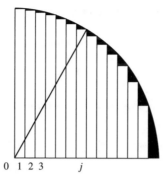

Figure 7.2. Approximation of the area of a quadrant.

when the symbol was introduced by William Jones (1675–1749). By around 2000 BC, the Babylonians used $3\frac{1}{8}$ as an approximation to π and the Egyptians were using $\pi \approx 4\left(\frac{8}{9}\right)^2$. There are many ways in which π can be obtained as a limit of a sequence or series. For example it is thought that Japanese mathematicians in the middle ages used arguments such as the following to estimate π.

Figure 7.2 shows a quadrant of a circle of radius 1 which is almost totally filled by n non-overlapping rectangles of equal width. The total area of the quadrant is $\frac{\pi}{4}$. Now let's calculate the area of the rectangles which approximates this. Each rectangular strip has width $\frac{1}{n}$ and by Pythagoras' theorem, the height of the jth strip is $\sqrt{1 - \frac{j^2}{n^2}}$. So we see that the area of the jth strip is

$$\frac{1}{n}\sqrt{1 - \frac{j^2}{n^2}} = \frac{1}{n^2}\sqrt{n^2 - j^2}.$$

Consequently we deduce that the total area of all the rectangles filling the quadrant is $\frac{1}{n^2}\sum_{j=1}^{n}\sqrt{n^2 - j^2}$. Now as n becomes larger and larger we can see that the rectangles fill up more and more of the quadrant and so it's reasonable to assert that

$$\frac{\pi}{4} = \lim_{n\to\infty}\frac{1}{n^2}\sum_{j=1}^{n}\sqrt{n^2 - j^2},$$

and so $\pi = 4\lim_{n\to\infty}\frac{1}{n^2}\sum_{j=1}^{n}\sqrt{n^2 - j^2}$. The reader should experiment with calculating approximations to π using this formula, e.g. what do you get when $n = 10, 50, 100$?

The invention of calculus enabled mathematicians to find more satisfying and direct expressions for π involving convergent series. We'll sketch a very famous argument but beware that there are lots of gaps that need filling – even for those who know how to do the calculus.

We begin with a well-known formula of integration:

$$\arctan(x) = \int_0^x \frac{1}{1+y^2}\,dy,$$

where $\arctan(x)$ denotes the angle in radians whose tan is x. By the binomial series (see Appendix 1) we have

$$\frac{1}{1+y^2} = 1 - y^2 + y^4 - y^6 + \cdots = \sum_{n=0}^{\infty}(-1)^n y^{2n},$$

provided $|y| \leq 1$. Integrating term-by-term we get *Gregory's series*

$$\arctan(x) = x - \frac{x^3}{3} + \frac{x^5}{5} - \frac{x^7}{7} + \cdots$$

Now if we pass to the limit as $x \to 1$ we finally find that since $\tan\left(\frac{\pi}{4}\right) = 1$ then

$$\frac{\pi}{4} = 1 - \frac{1}{3} + \frac{1}{5} - \frac{1}{7} + \cdots,$$

$$\pi = 4\left(1 - \frac{1}{3} + \frac{1}{5} - \frac{1}{7} + \cdots\right).$$

This formula is very beautiful but from the practical point of view of trying to calculate π it has a serious defect. It converges very slowly as you can find for yourself by calculating some partial sums. Now Gregory's series for π was first published in 1671. A series that converges much faster was obtained by Isaac Newton (and published in 1742). He followed a similar path to Gregory but instead of the integral for the inverse tangent, he used that for the inverse sine:

$$\arcsin(x) = \int_0^x \frac{1}{\sqrt{1-y^2}}\,dy.$$

Now expanding $(1-y^2)^{-\frac{1}{2}}$ as a binomial series and integrating term-by-term yields (for $-1 < x < 1$)

$$\arcsin(x) = x + \frac{1}{2}\frac{x^3}{3} + \frac{1.3}{2.4}\frac{x^5}{5} + \frac{1.3.5}{2.4.6}\frac{x^7}{7} \cdots$$

and using the fact that $\arcsin\left(\frac{1}{2}\right) = \frac{\pi}{6}$ we get

$$\pi = 6\left[\frac{1}{2} + \frac{1}{2.3}\left(\frac{1}{2}\right)^3 + \frac{1.3}{2.4.5}\left(\frac{1}{2}\right)^5 + \frac{1.3.5}{2.4.6.7}\left(\frac{1}{2}\right)^7 + \cdots\right]$$

and you should compute the first few terms of this and convince yourself that it really does appear to converge faster to π than Gregory's series does.

We have shown that e is irrational. Now we'll look at the corresponding result for π. As the proof is much harder I'm only going to sketch an outline. In fact the proof is slightly indirect in that we'll aim to show that π^2 is irrational. The irrationality of π itself then follows from I(iii) in Chapter 2.

Theorem 7.2.1. π^2 is an irrational number.

Proof. We proceed to give a proof by contradiction so let's assume that $\pi^2 = \frac{p}{q}$ where p and q are natural numbers. We'll split the proof into steps and start at what seems to be a highly irrelevant observation.

Step 1: A Curious Sideline. Fix a natural number n (which later on we will want to make large) and define the function

$$f(x) = \frac{1}{n!}x^n(1-x)^n \tag{7.2.7}$$

for all $0 \leq x \leq 1$. It's worth making a note of the inequality

$$f(x) < \frac{1}{n!}, \tag{7.2.8}$$

for all $0 \leq x \leq 1$.

We denote the first and second derivatives of f by f' and f'' and the kth derivative by $f^{(k)}$. Using the binomial theorem we have $(1-x)^n = \sum_{r=0}^{n} \binom{n}{r}(-x)^{n-r}$ where the binomial coefficients $\binom{n}{r} = \frac{n!}{(n-r)!r!}$ (see Appendix 1). It follows that

$$f(x) = \sum_{r=0}^{n} \frac{(-1)^{n-r}}{n!} \binom{n}{r} x^{2n-r},$$

and if we substitute $s = n - r$ we obtain

$$f(x) = \sum_{s=0}^{n} \frac{(-1)^s}{n!} \binom{n}{n-s} x^{n+s} = \sum_{s=0}^{n} \frac{c_s}{n!} x^{n+s}, \tag{7.2.9}$$

where $c_s = (-1)^s \binom{n}{n-s}$.

121

Now from (7.2.7) we have $f(x) = f(1 - x)$. From this it follows by repeated differentiation that $f^{(k)}(x) = (-1)^k f^{(k)}(1 - x)$. So in particular $f^{(k)}(1) = (-1)^k f^{(k)}(0)$. Now if $k < n$ or $k > 2n$, $f^{(k)}(0) = 0$. While if $n \leq k \leq 2n$ we see from (7.2.9) that

$$f^{(k)}(x) = \sum_{s=0}^{n-k} \frac{c_s}{n!}(n+s)(n+s-1)\cdots(n+s-k+1)x^{n+s-k}$$

and so

$$f^{(k)}(0) = \frac{c_{k-n}}{n!}k! = (-1)^{k-n}\binom{n}{2n-k}k(k-1)\cdots(n+1),$$

which is an integer. So we conclude that $f^{(k)}(0)$ (and hence $f^{(k)}(1)$) is an integer for all nonnegative integer values of k.

Step 2: Curiouser and Curiouser
Remember that q appears as the denominator of our conjectured rational number representation of π^2 and also don't forget our function f from (7.2.7). Now introduce a new function

$$g(x) = q^n(\pi^{2n}f(x) - \pi^{2n-2}f^{(2)}(x) + \pi^{2n-4}f^{(4)}(x) - \cdots + (-1)^n f^{(2n)}(x)). \tag{7.2.10}$$

If we use the facts that $f^{(k)}(0)$ and $f^{(k)}(1)$ are integers and $\pi^{2n} = \frac{p^n}{q^n}$, you can deduce that $g(0)$ and $g(1)$ are also both integers.

Differentiate (7.2.10) twice and rearrange to find that

$$g''(x) = -\pi^2 g(x) + q^n \pi^{2n+2} f(x). \tag{7.2.11}$$

Step 3: The Coup de Grace
We introduce yet a third mysterious function:

$$h(x) = g'(x)\sin(\pi x) - \pi g(x)\cos(\pi x).$$

Differentiation using the product rule and substitution from (7.2.11) yields

$$h'(x) = \pi^2 p^n f(x)\sin(\pi x). \tag{7.2.12}$$

Spitting $\pi^2 = \pi \times \pi$ and integrating both sides of (7.2.12) we get

$$\pi p^n \int_0^1 f(x)\sin(\pi x)dx = \frac{1}{\pi}\int_0^1 h'(x)dx$$

$$= \frac{1}{\pi}(h(1) - h(0))$$

$$= \frac{1}{\pi}[g'(1)\sin(\pi) - \pi g(1)\cos(\pi) - g'(0)\sin(0)$$

$$+ \pi g(0)\cos(0)]$$

$$= g(1) + g(0),$$

which is an integer. So we've deduced that $m(n) = \pi p^n \int_0^1 f(x)\sin(\pi x)dx$ is an integer for all natural numbers n. Furthermore, since both $f(x) > 0$ and $\sin(\pi x) > 0$ for $0 < x < 1$, it follows that m is in fact a natural number. Now from the series expansion for e^p we know that $\sum_{n=0}^{\infty} \frac{p^n}{n!}$ converges and so it follows from Theorem 6.4.1 that $\lim_{n\to\infty} \frac{p^n}{n!} = 0$. Hence by the definition of the limit we can find n sufficiently large so that $\frac{\pi p^n}{n!} < 1$. Using this fact together with the inequality (7.2.8) and the fact that $\sin(\pi x) \leq 1$ for $0 \leq x \leq 1$, we get for our (sufficiently large) value of n that

$$m(n) = \pi p^n \int_0^1 f(x)\sin(\pi x)dx < \pi p^n \int_0^1 \frac{1}{n!}dx = \frac{\pi p^n}{n!} < 1$$

and that gives the required contradiction. □

The fact that π is transcendental was proved by Ferdinand von Lindemann (1852–1939)[5] in 1882.[6] As we've mentioned transcendental numbers again, it's worth making a diversion to say something more about these. At the International Congress of Mathematicians that took place in Paris in 1900 the leading German mathematician David Hilbert (1862–1943) presented ten key problems for research in the twentieth century. He later published an article which included a further thirteen and the entire collection of twenty three have since become known as 'Hilbert problems'.[7] For his seventh problem, Hilbert asked if α^β is always transcendental if α is algebraic (but not equal to 0 or 1) and β is both algebraic and irrational. This was proved to be correct in 1934 by A.O.Gelfond (1906–68).[8] So for example all numbers of the form $p^{\sqrt{N}}$ are transcendental where p is prime and N is a natural number that is not a perfect square. A corollary of Gelfond's result (using complex numbers – see Chapter 8, section 3) is that e^π is also transcendental. However the status of the numbers π^e, e^e or π^π (i.e. whether they are algebraic or transcendental) remains an open problem.

7.3 The Number γ

The number γ is less well known than its famous cousins π and e but that doesn't make it any the less interesting. In this short section we'll briefly indicate where it

[5] See http://en.wikipedia.org/wiki/Ferdinand_von_Lindemann
[6] For those intrepid readers who want to see proofs that both e and π are transcendental, see e.g. pages 214–5 and 217–8 (respectively) of J.K. Truss *Foundations of Mathematical Analysis*, Oxford University Press (1997) – but be aware that this is a graduate level text.
[7] See http://en.wikipedia.org/wiki/Hilbert's_problems
[8] See http://en.wikipedia.org/wiki/Aleksandr_Gelfond. The result was also proved independently in 1935 by Theodor Schneider (1911–88) (http://en.wikipedia.org/wiki/Theodor_Schneider) in his PhD thesis.

comes from. We'll start with the divergent series $\sum_{n=1}^{\infty} \frac{1}{n}$. Integrals are continuous versions of sums so a close relative to this series is the divergent integral $\int_1^{\infty} \frac{1}{x} dx$. Leonhard Euler had the brilliant idea of investigating the sequence (b_N) whose Nth term is $\sum_{n=1}^{N} \frac{1}{n} - \int_1^N \frac{1}{x} dx$. But $\int_1^N \frac{1}{x} dx = \log(N)$ so

$$b_N = \sum_{n=1}^{N} \frac{1}{n} - \log(N).$$

Theorem 7.3.1. The sequence (b_N) converges.

Proof. Since

$$\log(N) = \int_1^N \frac{1}{x} dx = \sum_{n=1}^{N-1} \int_n^{n+1} \frac{1}{x} dx$$

and

$$\sum_{n=1}^{N} \frac{1}{n} = \frac{1}{N} + \sum_{n=1}^{N-1} \frac{1}{n} = \frac{1}{N} + \sum_{n=1}^{N-1} \int_n^{n+1} \frac{1}{n} dx,$$

we have

$$b_N = \sum_{n=1}^{N-1} \int_n^{n+1} \left(\frac{1}{n} - \frac{1}{x} \right) dx + \frac{1}{N}.$$

Now whenever $n \leq x \leq n+1$,

$$0 \leq \frac{1}{n} - \frac{1}{x} = \frac{x-n}{nx} \leq \frac{1}{n^2} \qquad \ldots \text{(i)}$$

and so we have

$$0 \leq b_N = \sum_{n=1}^{N-1} a_n + \frac{1}{N},$$

where $a_n = \int_n^{n+1} \left(\frac{1}{n} - \frac{1}{x} \right) dx$.

It follows from (i) that $\sum_{n=1}^{N-1} a_n \leq \sum_{n=1}^{N-1} \frac{1}{n^2}$. By the comparison test $\sum_{n=1}^{\infty} a_n$ converges and since $\lim_{N \to \infty} \frac{1}{N} = 0$, we deduce by algebra of limits that the sequence (b_N) converges. $\square$

We write $\gamma = \lim_{N \to \infty} b_N$. The decimal expansion of γ begins $0.577215664\ldots$ and it is sometimes called the Euler–Mascheroni constant.[9] It is an unsolved problem to determine whether γ is rational or irrational.

γ is a truly amazing number and to find out why, I recommend the superb book by Julian Havill 'Gamma: Exploring Euler's Constant' (see also the Further Reading section). Here you will learn about the fascinating relationship between the number γ and the Gamma function that is defined by $\Gamma(y) = \int_0^\infty x^{y-1} e^{-x} dx$. Using integration by parts, it's not difficult to show that $\Gamma(n+1) = n!$ for all natural numbers n, so Γ can be thought of as a generalisation of the factorial to more general real numbers. It's turns out that Γ is differentiable and if $\Gamma'(y)$ denotes the value of its derivative at the point y then

$$\gamma = -\Gamma'(1),$$

which is truly remarkable!

---

[9] The Italian mathematician Lorenzo Mascheroni (1750–1800) accurately calculated the first nineteen numbers in the decimal expansion of γ.

Infinite Products

However the man got a little bit excited; he wanted to prove himself some more. "Multiplicação!" he said.

Surely You're Joking, Mr Feynman, R.P. Feynman

8.1 Convergence of Infinite Products

Just as we considered infinite sums or series of numbers, we can also deal with infinite products of numbers as limits. To be precise, suppose that we are given a sequence of real numbers (a_n). We can consider the associated sequence of *partial products* (p_n) where $p_n = a_1 a_2 \cdots a_n$ and enquire whether this sequence converges to a limit. Just as we used the sigma notation for sums (the Greek letter S), we use the Greek letter Π for P when we deal with products. So we write

$$p_n = \prod_{i=1}^{n} a_i,$$

and if $p = \lim_{n \to \infty} p_n$ exists then we write

$$p = \prod_{i=1}^{\infty} a_i,$$

so that $\prod_{i=1}^{\infty} a_i = \lim_{n \to \infty} \prod_{i=1}^{n} a_i$. Of course for the limit to exist we require as usual that for any $\epsilon > 0$ there exists a natural number N so that $|p_n - p| < \epsilon$ whenever $n > N$.

Here are some rather easy assertions that you may want to try to prove for yourself:

(a) If $a_n = 0$ for some value of n then $p_m = 0$ for all $m \geq n$ and $p = 0$.

126

(b) If $a_n = c$ for all n, where c is a non-zero constant then (p_n) diverges if $c \leq -1$ and $c > 1$, converges to 1 when $c = 1$ and to 0 when $|c| < 1$.

(c) If $0 < a_n < 1$ and (a_n) is monotonic decreasing then (p_n) converges to 0. (Hint: Note that $p_n \leq a_1^n$.)

So for example as a consequence of (c), you can easily deduce that $\prod_{n=1}^{\infty} \frac{1}{n+1}$ converges to 0 and hence so does $\prod_{n=1}^{\infty} \frac{1}{n} = 1 \times \prod_{n=1}^{\infty} \frac{1}{n+1}$.

The mathematics literature contains far less about infinite products than it does about infinite sums. This may be because (to a large extent) the study of convergence of products can be reduced to that of sums. For example if each $a_n > 0$ and you know about logarithms,[1] it can be shown that

$$\log\left(\prod_{i=1}^{\infty} a_i\right) = \sum_{i=1}^{\infty} \log(a_i), \tag{8.1.1}$$

and so

$$\prod_{i=1}^{\infty} a_i = \exp\left(\sum_{i=1}^{\infty} \log(a_i)\right).$$

I'm not really assuming much knowledge of logarithms in this book and we won't use them again in this chapter. The next inequality gives a nice link between convergence of certain infinite series and that of related infinite products. However the proof uses some facts about continuous functions that aren't dealt with in this book. Rest assured that you can skip the proof if you want to. We won't use the result anywhere else in the book.

Theorem 8.1.1. If (a_n) is such that each $a_n = 1 + \alpha_n$ where $\alpha_n > 0$, then $\sum_{i=1}^{\infty} \alpha_i$ converges if and only if $\prod_{i=1}^{\infty} a_i$ converges and in either case we have

$$\sum_{i=1}^{\infty} \alpha_i \leq \prod_{i=1}^{\infty} a_i \leq \exp\left(\sum_{i=1}^{\infty} \alpha_i\right).$$

Proof. We'll begin by proving the inequality for finite sums and products, i.e. when ∞ is replaced by n throughout. First consider the left-hand inequality. It is

---

[1] Here I'm using the notation 'log' to mean 'log$_e$' i.e. the logarithm to base e, so that $e^{\log(x)} = \log(e^x)$ for all $x > 0$. Note that some textbooks denote log by ln.

fairly easy to verify this as

$$\prod_{i=1}^{n}(1+\alpha_i) = (1+\alpha_1)(1+\alpha_2)\cdots(1+\alpha_n)$$

$$\geq \sum_{i=1}^{n}\alpha_i,$$

as $\sum_{i=1}^{n}\alpha_i$ is just one of the terms that we get when we expand out the brackets on the right-hand side.

For the right-hand inequality, by (7.1.6) $e^{\alpha_i} \geq 1+\alpha_i$ for each i and so

$$\prod_{i=1}^{n}(1+\alpha_i) \leq e^{\alpha_1}e^{\alpha_2}\cdots e^{\alpha_n} = \exp\left(\sum_{i=1}^{n}\alpha_i\right).$$

So we have shown that for all natural numbers n

$$\sum_{i=1}^{n}\alpha_i \leq \prod_{i=1}^{n}a_i \leq \exp\left(\sum_{i=1}^{n}\alpha_i\right). \tag{8.1.2}$$

Now suppose that $\sum_{i=1}^{\infty}\alpha_i$ converges then the sequence whose nth term is $\exp\left(\sum_{i=1}^{n}\alpha_i\right)$ converges to $\exp\left(\sum_{i=1}^{\infty}\alpha_i\right)$. To verify this you need the fact that the function $x \to e^x$ is *continuous* which enables us to legitimately conclude that

$$\lim_{n\to\infty}\exp\left(\sum_{i=1}^{n}\alpha_i\right) = \exp\left(\sum_{i=1}^{\infty}\alpha_i\right).$$

The study of continuous functions lies beyond the scope of this book – but it can be found in any standard introductory text on analysis (see Chapter 12 for a brief taster). It then follows from (8.1.2),[2] that for all n

$$\prod_{i=1}^{n}a_i \leq \exp\left(\sum_{i=1}^{\infty}\alpha_i\right),$$

and the convergence of the infinite product may now be deduced by using Theorem 5.2.1. The proof of the rest of the theorem is left as an exercise for the reader. □

When we studied infinite series we had the interesting result (Theorem 6.4.1) that if $\sum_{n=1}^{\infty}a_n$ converges then $\lim_{n\to\infty}a_n = 0$. There is an analogous result for infinite products and here it is:

[2] You also need the fact that $e^x \leq e^y$ if $x \leq y$.

Theorem 8.1.2. Let (a_n) be a sequence with each $a_n > 0$. If $\prod_{i=1}^{\infty} a_i$ converges to a limit l and $l \neq 0$ then (a_n) converges and $\lim_{n \to \infty} a_n = 1$.

Proof. This result can be deduced very quickly from (8.1.1) and Theorem 6.4.1 or (in the case where each $a_i > 1$) from Theorems 8.1.1 and 6.4.1. We'll leave the details of this to the reader. Below I'll give a 'stand-alone' proof that requires no knowledge of logarithms or the constraint that $a_i > 1$.

By definition of a limit, we know that given any $\epsilon > 0$ there exists a natural number N such that if $n > N$ then $\left| \prod_{i=1}^{n} a_i - l \right| < \epsilon$, i.e.

$$l - \epsilon < \prod_{i=1}^{n} a_i < l + \epsilon.$$

Now using (L5) we have that

$$\frac{l - \epsilon}{l + \epsilon} < a_{n+1} = \frac{\prod_{i=1}^{n+1} a_i}{\prod_{i=1}^{n} a_i} < \frac{l + \epsilon}{l - \epsilon}.$$

i.e. $$1 - \frac{2\epsilon}{l + \epsilon} < a_{n+1} < 1 + \frac{2\epsilon}{l - \epsilon}.$$

By making use of (L4) and (L5) you can check that $-\frac{1}{l-\epsilon} < -\frac{1}{l+\epsilon}$ and so

$$1 - \frac{2\epsilon}{l - \epsilon} < a_{n+1} < 1 + \frac{2\epsilon}{l - \epsilon}.$$

In other words

$$|a_{n+1} - 1| < \frac{2\epsilon}{l - \epsilon}.$$

Now remember this holds for all $n > N$ and observe that $\frac{2\epsilon}{l-\epsilon}$ can be made arbitrarily small. Indeed pick $\delta > 0$ to be as small as you like. Then to get $\frac{2\epsilon}{l-\epsilon} < \delta$ is equivalent to ensuring that $\epsilon < \frac{\delta l}{2+\delta}$ which you are at perfect liberty to do. The result then follows. $\qquad \square$

There is a beautiful way of obtaining π as an infinite product which was first discovered by the English mathematician John Wallis (1616–1703) and published in 1655 in his *Arithmetica Infinitorum*.[3] His formula is usually written:

$$\frac{\pi}{2} = \frac{2^2 4^2 6^2 \cdots}{3^2 5^2 7^2 \cdots},$$

but in the light of the work we've done in this chapter, we can write

$$\frac{\pi}{2} = \prod_{n=1}^{\infty} \frac{(2n)^2}{(2n+1)^2}.$$

To see why this result is true you'll need to master the calculus.

---

[3] See e.g. http://en.wikipedia.org/wiki/John_Wallis

8.2 Infinite Products and Prime Numbers

In this section we will aim to understand a beautiful formula which links infinite series of negative powers of natural numbers with infinite products involving prime numbers. This result was discovered by Leonhard Euler whose work we have already encountered several times in this book. Not only does it constitute a piece of wonderful mathematics in itself, but it was also the launching point for some marvellous work by Bernard Riemann (1826–66) which led to what is now called the *Riemann zeta function*.

Consider the infinite series $\sum\limits_{n=1}^{\infty} \frac{1}{n^r}$ where $r > 1$. In honour of Riemann's later discovery we'll write

$$\zeta(r) = \sum_{n=1}^{\infty} \frac{1}{n^r},$$

where ζ is the Greek letter 'zeta' which plays the same role as the English 'z' and is pronounced 'zeetah'.

Euler's great result is:

Theorem 8.2.1. For all $r > 1$,

$$\sum_{n=1}^{\infty} \frac{1}{n^r} = \prod_{p} \left(1 - \frac{1}{p^r}\right)^{-1}. \tag{8.2.3}$$

Note that here $\prod_p$ means the product over all the prime numbers so

$$\prod_{p} \left(1 - \frac{1}{p^r}\right)^{-1} = \left(1 - \frac{1}{2^r}\right)^{-1} \left(1 - \frac{1}{3^r}\right)^{-1} \left(1 - \frac{1}{5^r}\right)^{-1} \left(1 - \frac{1}{7^r}\right)^{-1} \cdots$$

$$= \frac{2^r 3^r 5^r 7^r 11^r \cdots}{(2^r - 1)(3^r - 1)(5^r - 1)(7^r - 1)(11^r - 1) \cdots}.$$

We'll give two outline proofs of Theorem 8.2.3:

Proof 1. For each prime number p, $\frac{1}{p} < 1$, and so by a binomial series expansion

$$\left(1 - \frac{1}{p^r}\right)^{-1} = 1 + \frac{1}{p^r} + \frac{1}{p^{2r}} + \frac{1}{p^{3r}} + \cdots. \tag{8.2.4}$$

Now fix a natural number N and consider $\prod_{p<N} \left(1 - \frac{1}{p^r}\right)^{-1}$. When we collect together all the terms obtained when all the different series of the form (8.2.4) are multiplied together, we obtain an infinite sum of all possible terms of the form

130

$\frac{1}{p_1^{n_1} p_2^{n_2} \cdots p_k^{n_k}}$ where $p_1, p_2, \ldots, p_k$ are prime numbers less than N and $n_1, n_2, \ldots, n_k$ are natural numbers. It follows by Theorem 1.2.1 that

$$\prod_{p<N} \left(1 - \frac{1}{p^r}\right)^{-1} = \sum \frac{1}{n^r},$$

where the sum is over all natural numbers whose prime factors are no larger than N. Now let $N \to \infty$ and the result follows.[4]   □

Proof 2. By definition

$$\zeta(r) = 1 + \frac{1}{2^r} + \frac{1}{3^r} + \frac{1}{4^r} + \frac{1}{5^r} + \cdots$$

Then

$$\left(1 - \frac{1}{2^r}\right) \zeta(r) = \zeta(r) - \frac{1}{2^r}\zeta(r)$$

$$= \left(1 + \frac{1}{2^r} + \frac{1}{3^r} + \frac{1}{4^r} + \frac{1}{5^r} + \cdots\right) - \left(\frac{1}{2^r} + \frac{1}{4^r} + \frac{1}{6^r} + \frac{1}{8^r}\right) \cdots$$

$$= 1 + \frac{1}{3^r} + \frac{1}{5^r} + \frac{1}{7^r} + \frac{1}{9^r} + \cdots$$

We next compute

$$\left(1 - \frac{1}{3^r}\right)\left(1 - \frac{1}{2^r}\right) \zeta(r) = \left(1 + \frac{1}{3^r} + \frac{1}{5^r} + \frac{1}{7^r} + \frac{1}{9^r} + \cdots\right)$$

$$- \left(\frac{1}{3^r} + \frac{1}{9^r} + \frac{1}{15^r} + \frac{1}{21^r} + \cdots\right)$$

$$= 1 + \frac{1}{5^r} + \frac{1}{7^r} + \frac{1}{11^r} + \frac{1}{13^r} + \frac{1}{17^r} \cdots$$

Now let's take stock. Multiplying $\zeta(r)$ by $\left(1 - \frac{1}{2^r}\right)$ removed all powers of $\frac{1}{2}$. A further multiplication by $\left(1 - \frac{1}{3^r}\right)$ removed all powers of $\frac{1}{3}$. If we continue multiplying up to some large prime p then $\left(1 - \frac{1}{p^r}\right) \cdots \left(1 - \frac{1}{3^r}\right)\left(1 - \frac{1}{2^r}\right)\eta(r) - 1$ is a sum of terms of the form $\frac{1}{n^r}$ where n has no prime factor p or smaller. Now let $p \to \infty$ and deduce that in the limit this sum converges to zero. The result follows.[5]   □

---

[4] You can and should question whether this argument really deserves to be called a proof.
[5] This isn't precise enough – can you make it so?

As a direct consequence of Theorem 8.2.3 we may write

$$\zeta(r)^{-1} = \prod_{p} \left(1 - \frac{1}{p^r}\right).$$

If we expand out the product on the right-hand side (and just for once, let's not worry too much about rigour) we get

$$\zeta(r)^{-1} = 1 - \sum_{p} \frac{1}{p^r} + \sum_{p_1 < p_2} \frac{1}{(p_1 p_2)^r} - \sum_{p_1 < p_2 < p_3} \frac{1}{(p_1 p_2 p_3)^r} + \cdots$$

Notice that the right-hand side is a sum of reciprocals of integers, written in terms of their prime factorisation, in which no square appears (as every prime factor appears only once). In fact we can write this succinctly as follows:

$$\zeta(r)^{-1} = \sum_{n=1}^{\infty} \frac{\mu(n)}{n^r},$$

where μ is the *Mobius function*,[6] which is defined as follows:

$$\mu(n) = \begin{cases} 0 & \text{if } n \text{ fails to be square-free,} \\ 1 & \text{if } n = 1 \text{ or is the product of an even number of distinct primes.} \\ -1 & \text{if } n \text{ is the product of an odd number of distinct primes.} \end{cases}$$

The Mobius function plays an important role in number theory but we will give no more than this brief introduction to it here.

The next two results use only finite (rather than infinite) products but this seems like an ideal opportunity to include them. First let us return to Chapter 1. Recall Theorem 1.2.2 where we gave the classic proof by Euclid that there are infinitely many primes. More than 2000 years later, Leonhard Euler gave a different and very elegant proof that employs infinite series.

Theorem 8.2.2 (Theorem 1.2.2 revisited). There are an infinite number of prime numbers.

Proof. We again use proof by contradiction and assume that there are only a finite number of primes, N, say, and we write them in order as $p_1, p_2, \ldots, p_N$ (of course $p_1 = 2$ and $p_2 = 3$ but the notation is convenient). We will use the fact (which we proved in Theorem 1.2.1) that every natural number n has a prime factorisation $n = p_1^{m_1} p_2^{m_2} \cdots p_N^{m_N}$ (where some of the m_is may be zero). Now recall the sequence (s_n) of partial sums of the harmonic series: $s_n = 1 + \frac{1}{2} + \cdots + \frac{1}{n}$ and for now let us fix a value of n. Let $m = \max\{m_1, m_2, \ldots, m_N\}$ and consider the

---

[6] Named after the German mathematician Augustus Ferdinand Mobius (1790–1868).

product

$$\left(1 + \frac{1}{2} + \frac{1}{2^2} + \cdots + \frac{1}{2^m}\right)\left(1 + \frac{1}{3} + \frac{1}{3^2} + \cdots + \frac{1}{3^m}\right)\cdots$$

$$\cdots \left(1 + \frac{1}{p_N} + \frac{1}{p_N^2} + \cdots + \frac{1}{p_N^m}\right),$$

which we can write succinctly as $\prod_{i=1}^{N} \sum_{k=0}^{m} \frac{1}{p_i^k}$. The key observation is that

$$s_n \leq \prod_{i=1}^{N} \sum_{k=0}^{m} \frac{1}{p_i^k},$$

since every natural number whose reciprocal appears in the sum on the left-hand side has a prime factorisation whose reciprocal appears on the right-hand side. But for each $1 \leq i \leq N$, using the formula for the sum of a geometric series, we have

$$\sum_{k=0}^{m} \frac{1}{p_i^k} \leq \sum_{k=0}^{\infty} \frac{1}{p_i^k} = \frac{1}{1 - \frac{1}{p_i}} = \frac{p_i}{1 - p_i},$$

and so we conclude that

$$s_n \leq \prod_{i=1}^{N} \frac{p_i}{1 - p_i}.$$

Now the term on the right-hand side is completely independent of the choice of n and so we conclude that the sequence (s_n) is bounded above. But it is also monotonic increasing and so it has a limit. This contradicts the known fact that the harmonic series diverges and enables us to conclude that there cannot be a finite number of prime numbers. □

To close this section, we'll prove a result that uses both the properties of the exponential function that were developed in the last chapter, and the notation for finite products that we've seen in this one. We've already seen that $\sum_{i=1}^{\infty} \frac{1}{i}$ diverges and that even $\sum_{i=1}^{\infty} \frac{1}{i_{sf}}$ diverges where the sum is over square-free integers only. Of course every prime number is square-free and you may have been wondering about $\sum_{p=1}^{\infty} \frac{1}{p}$ where the sum is over all prime numbers. The question was (again) settled by Leonhard Euler.[7]

---

[7] The proof given here is due to Ivan Niven *The American Mathematical Monthly* Vol. **78**, No. 2 pp. 272–3 (1971) and I found it on pages 75–6 of 'Euler The Master of Us All' by William Dunham, Math. Assoc. of America (1999).

Theorem 8.2.3. $\sum\limits_{p=1}^{\infty} \frac{1}{p}$ diverges.

Proof. We'll give a proof by contradiction. Suppose that $\sum\limits_{p=1}^{\infty} \frac{1}{p}$ converges to a real number L. Fix a natural number n and let q be the largest prime number that doesn't exceed n. Now since L is the supremum of the sequence of partial sums we have (using (ii) just after (7.1.6)) that

$$e^L > \exp\left\{ \frac{1}{2} + \frac{1}{3} + \frac{1}{5} + \cdots + \frac{1}{q} \right\}$$

$$= \prod_{2 \leq p \leq q} e^{\frac{1}{p}}$$

$$\geq \prod_{2 \leq p \leq q} \left(1 + \frac{1}{p} \right) \quad \text{by (7.1.6)}$$

$$= \left(1 + \frac{1}{2} \right)\left(1 + \frac{1}{3} \right) \cdots \left(1 + \frac{1}{q} \right)$$

$$\geq \sum_{i_{sf} \leq n} \frac{1}{i_{sf}},$$

where the sum is over all square-free integers less than or equal to n. This inequality is correct since the square-free integers are precisely those that have no squares (or higher powers) of primes in their prime factorisation and we get the reciprocal of all those that don't exceed n (and also some that do) when we multiply out the bracket. So we've shown that e^L is an upper bound for the monotonic increasing sequence whose nth term is $\sum\limits_{i_{sf} \leq n} \frac{1}{i_{sf}}$ and so this sequence converges by Theorem 5.2.1(i) and that contradicts Theorem 6.5.2. Hence we deduce that $\sum\limits_{p=1}^{\infty} \frac{1}{p}$ diverges. $\qquad\qquad\square$

8.3 Diversion – Complex Numbers and the Riemann Hypothesis

One of the greatest achievements of Bernard Riemann was to extend $\zeta(r)$ so that the real number r is replaced by a *complex number*. Complex numbers first appeared in the work of Rafael Bombelli (1526–72),[8] who instead of treating

---

[8] See e.g. http://en.wikipedia.org/wiki/Rafael_Bombelli

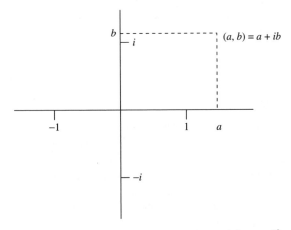

Figure 8.1. Complex numbers in the 'Argand diagram'.[9]

$x^2 = -1$ as an equation that couldn't be solved, decided to pretend that it could. As this is a quadratic equation, it must have two roots and since these cannot be real numbers, let's call them i and $-i$. So i is a new type of number that has the property that $i^2 = -1$. In fact we can give i a nice geometric interpretation in the dynamic spirit of Section 1.3 where we viewed negative numbers as reflections through an imaginary mirror on the y-axis. We should again think of the real line as sitting inside an infinite two-dimensional plane as in Figure 1.3. Recall that a reflection is the same thing as a rotation (and for convenience, we'll take this to be anti-clockwise) through 180 degrees. Now let's suppose that we instead rotate (again anti-clockwise) through 90 degrees. We can identify this rotation with the new number i, so i takes the point with co-ordinates $(1, 0)$ on the x-axis to the point with co-ordinates $(0, 1)$ on the y-axis. If we then make another 90 degree rotation, we will have rotated 180 degrees in total and reach the number $(-1, 0)$. So i^2 takes us from $(1, 0)$ to $(-1, 0)$, i.e. $i^2 = -1$.

When we think in this way, any point in the plane which has co-ordinates (a, b) (where a and b are real numbers) is re-interpreted as a complex number $a + ib$ (see Figure 8.1). Such numbers can be added by the rule:

$$(a + ib) + (c + id) = (a + c) + i(b + d),$$

and also multiplied together by expanding brackets in the usual way, remembering that $i^2 = -1$ so

$$(a + ib)(c + id) = ab + ibc + iad + i^2 bd$$

$$= (ac - bd) + i(bc + ad).$$

---

[9] The representation of complex numbers as points in two-dimensional space is often called an Argand diagam after Jean-Robert Argand (1768–1822).

We can even discuss convergence of sequences of complex numbers, so if (z_n) is such a sequence where each $z_n = a_n + ib_n$ we say that (z_n) converges whenever both of the real-valued sequences (a_n) and (b_n) do. Indeed in this case, if (a_n) converges to a and (b_n) converges to b then (z_n) converges to $a + ib$, e.g. $\lim_{n\to\infty} \left\{ \frac{1}{n} + i\left(1 - \frac{1}{n}\right) \right\} = i$. This also enables us to develop a theory of infinite series $\sum_{n=1}^{\infty} z_n$ where each z_n is a complex number. All of this is much more than just a delightful game. Complex numbers have a rich and powerful theory that sometimes throws great light on facts about real numbers.[10] Indeed Jaques Hadamard who we met earlier through his work on the prime number theorem wrote that 'The shortest path between two truths in the real domain sometimes passes through the complex domain'. As an example of the unifying power of complex numbers we'll mention the *fundamental theorem of algebra* which states that any algebraic equation of the form

$$a_0 + a_1 x + a_2 x^2 + \cdots + a_n x^n = 0,$$

where $a_0, a_1, a_2, \ldots, a_n$ are real numbers has n solutions, once complex numbers are allowed to come in and play. Complex numbers also play a fundamental role in many applications of mathematics to science and engineering, indeed they are a vital tool in the quantum theory which underlies our understanding of molecules, atoms and the microscopic realms of elementary particles.

Now let's return to Riemann and the zeta function. He considered the function

$$\zeta(z) = \sum_{n=1}^{\infty} \frac{1}{n^{x+iy}}, \qquad (8.3.5)$$

where $z = x + iy$. This series only converges when $x > 1$ but Riemann realised that it makes sense to 'analytically continue' the function ζ into the complex plane for the case where $x \leq 1$. We should be clear that ζ looks different for $x \leq 1$ than it does for $x > 1$ – indeed if we naively substitute $x = -1$ and $y = 0$ into (8.3.5) we get the divergent series $\sum_{n=1}^{\infty} n$. There is something much more subtle going on here. When thinking of ζ for $x < 1$ it is better to try to imagine a completely different function (one for which there is no clean formula) which naturally flows into (8.3.5) as we pass the $x = 1$ threshold.

Riemann focussed attention on the equation

$$\zeta(z) = 0.$$

It turns out that this equation has an infinite number of solutions that lie in the so-called critical strip, $0 \leq x \leq 1$. The celebrated *Riemann hypothesis* is that all of these are such that $x = \frac{1}{2}$. This problem has remained unsolved for over 150 years and may be the most important mathematical problem of all time. Its

[10] Any real number x can be identified with the complex number $x + i0$.

solution will lead to important information about prime numbers. Let's be a little more precise about this. Recall the prime number theorem that was discussed at the end of Section 4.1. It tells us that $\lim_{n \to \infty} \frac{\pi(n) \log_e(n)}{n} = 1$ where $\pi(n)$ is the number of prime numbers which are less than or equal to n. In fact an even better approximation to $\pi(n)$ can be found by using the logarithmic integral defined by $\text{li}(x) = \int_2^x \frac{1}{\log_e(y)} dy$ for $x > 2$, and we have the stronger form of the prime number theorem: $\lim_{n \to \infty} \frac{\pi(n)}{\text{li}(n)} = 1$. Now if the Riemann hypothesis is valid then we would get very precise information about how good an approximation $\text{li}(n)$ is to $\pi(n)$. In fact it can be shown that if the Riemann hypothesis is true then for all $n > 2657$,[11]

$$|\pi(n) - \text{li}(n)| \le \frac{1}{8\pi} \sqrt{n} \log_e(n). \qquad (8.3.6)$$

If we divide both sides of (8.3.6) by $\text{li}(n)$ we get

$$\left| \frac{\pi(n)}{\text{li}(n)} - 1 \right| \le \frac{1}{8\pi} \frac{\sqrt{n} \log_e(n)}{\text{li}(n)}$$

and since $\lim_{n \to \infty} \frac{\sqrt{n} \log_e(n)}{\text{li}(n)} = 0$, we see that (8.3.6) yields an 'error estimate' giving information on how close an approximation $\text{li}(n)$ is to $\pi(n)$ for very large n.

At the time of writing, it has been shown that the first 100 billion zeroes of ζ do indeed satisfy $x = \frac{1}{2}$ but of course, the conjecture may still be false. There are a number of popular accounts of the Riemann hypothesis, see e.g. *Prime Obsession* by John Derbyshire (Penguin 2004), *Dr Riemann's Zeros* by Karl Sabbagh (Atlantic Books 2002) and *The Music of the Primes* by Marcus de Sautoy (Fourth Estate 2003). The solution of the Riemann hypothesis featured within the eighth of Hilbert's 23 problems that were mentioned at the end of Section 7.2. One hundred years later it remained as one of seven unsolved 'Millennium problems'[12] that were announced by the Clay Mathematics Institute in May 2000.

---

[11] This result was established by the American mathematician Lowell Schoenfeld (1920–2002) in 1976.

[12] See http://www.claymath.org/millennium/

Continued Fractions

Continued fractions are part of the "lost mathematics", the mathematics now considered too advanced for high school and too elementary for college.

A History of π, P. Beckmann

In this chapter we'll take a brief look at an alternative and attractive way of representing real numbers. First we'll need a preliminary result that is of great interest in its own right.

9.1 Euclid's Algorithm

Recall that the highest common factor (hcf) of two natural numbers is the largest natural number that divides them both and leaves no remainder. It is also called the greatest common divisor (gcd). We'll write hcf(x, y) for the highest common factor of x and y, so e.g. hcf(3, 9) = 3, hcf(24, 108) = 12 and hcf(4, 7) = 1. Numbers such as 4 and 7 whose highest common factor is 1 are said to be *coprime*. Note that being coprime is a relative property of a pair of numbers and is nothing to do with either of the numbers being prime – e.g. 8 and 9 are coprime though neither is prime. Another useful bit of notation is $x|y$ which means that x divides into y and leaves no remainder – so we have for example, 4|96, 7|245 and 19|171.

Euclid gave a practical algorithm for finding hcf(x, y) and we will now describe how this works. Suppose that $x < y$ and divide x into y to get

$$y = cx + d,$$

where c is the unique natural number for which $cx \leq y < (c + 1)x$ (from which it follows that $d < x$). If $d = 0$ we are finished as y is then a multiple of x and so hcf$(x, y) = x$. So from now on we'll assume that $d \neq 0$. Now notice that any factor of both x and y is also a factor of d since $d = y - cx$. Also any factor of both x and d is also a factor of y. It follows that hcf$(y, x) = $ hcf(x, d) and we might as well replace y and x by the smaller pair of numbers x and d. From now on we'll write c as c_1 and d as d_1 as we are going to iterate the above process. So since $d_1 < x$ we can divide d_1 into x to get

$$x = c_2 d_1 + d_2.$$

Arguing above we have hcf$(x, d_1) = $ hcf(d_1, d_2) and so we can replace x and d_1 by d_1 and d_2. Continuing in this manner we generate a decreasing sequence of natural numbers $d_1, d_2, d_3, \ldots$ and since these are whole numbers the sequence must terminate at some point N, i.e.

$$d_{N-1} = c_{N+1} d_N. \qquad \text{(i)}$$

Going back one step we have

$$d_{N-2} = c_N d_{N-1} + d_N, \qquad \text{(ii)}$$

so any factor of both d_{N-2} and d_{N-1} is a factor of d_N. Arguing backwards we thus deduce that any factor of both x and y is a factor of d_N. But d_N is a factor of d_{N-1} by (i) and hence by (ii) is also a factor of d_{N-2}. Again arguing backwards we deduce that d_N is a factor of both x and y. It follows that $d_N = $ hcf(x, y).

An example will help us see what's going on. Let $y = 93$ and $x = 36$. We write

$$93 = (2 \times 36) + 21$$
$$36 = (1 \times 21) + 15$$
$$21 = (1 \times 15) + 6$$
$$15 = (2 \times 6) + 3$$
$$6 = 2 \times 3$$

So hcf$(93, 36) = 3$. In this case we have $N = 4$, $c_1 = 2$, $c_2 = 1$, $c_3 = 1$, $c_4 = 2$, $c_5 = 2$, $d_1 = 21$, $d_2 = 15$, $d_3 = 6$ and $d_4 = 3$.

9.2 Rational and Irrational Numbers as Continued Fractions

We'll carry on with the example we've just presented and rewrite it in fractional form:

$$\frac{93}{36} = 2 + \frac{21}{36} \qquad \ldots \text{(i)}$$
$$\frac{36}{21} = 1 + \frac{15}{21} \qquad \ldots \text{(ii)}$$

$$\frac{21}{15} = 1 + \frac{6}{15} \qquad \ldots \text{(iii)}$$

$$\frac{15}{6} = 2 + \frac{3}{6} \qquad \ldots \text{(iv)}$$

$$\frac{6}{3} = 2 \qquad \ldots \text{(v)}$$

Now substitute (ii) into (i) to get

$$\frac{93}{36} = 2 + \frac{1}{1 + \dfrac{15}{21}}.$$

Substituting from (iii) we obtain

$$\frac{93}{36} = 2 + \frac{1}{1 + \dfrac{1}{1 + \dfrac{6}{15}}}.$$

Now (iv) allows us to change this to

$$\frac{93}{36} = 2 + \frac{1}{1 + \dfrac{1}{1 + \dfrac{1}{2 + \dfrac{3}{6}}}},$$

and finally we incorporate (v) to conclude that

$$\frac{93}{36} = 2 + \frac{1}{1 + \dfrac{1}{1 + \dfrac{1}{2 + \dfrac{1}{2}}}}.$$

This is the *continued fraction* representation of the number $\frac{93}{36}$. We are going to investigate which other numbers we can write in this form so let's introduce some notation. Let (q_n) be a sequence of nonnegative integers. We'll call

$$f = q_0 + \cfrac{1}{q_1 + \cfrac{1}{q_2 + \cfrac{1}{q_3 + \cfrac{1}{q_4 + \cfrac{1}{q_5 + \cdots}}}}} \qquad (9.2.1)$$

a *regular continued fraction*. We'll say the fraction f *terminates* if $q_{N+1} = 0$ for some N. In this case f is perfectly well defined. If f doesn't terminate then it is not clear at this stage what meaning can be given to (9.2.1) and we'll come back to this point. You may guess that it will have something to do with limits. (9.2.1) is a pain to write out so we'll need some alternative notation. As a simplified representation of (9.2.1) we will write

$$f = [q_0, q_1, q_2, q_3, q_4, \ldots],$$

and if f terminates at $N > 2$ then $f = [q_0, q_1, q_2, \ldots, q_N]$. For example if we take $f = \frac{93}{36}$ then $N = 4$ and $f = [2, 1, 1, 2, 2]$. Note that, in general, we have the useful identity:

$$[q_0, q_1, q_2, q_3, q_4, \ldots] = q_0 + \cfrac{1}{[q_1, q_2, q_3, q_4, \ldots]}$$

$$= [q_0, [q_1, q_2, q_3, q_4, \ldots]].$$

Most of our work in this short chapter will be concerned with the regular case but I will also want to show you some examples of *irregular continued fractions*. For these the numerators in the continued fractions may differ from 1 so, as well as the sequence (q_n), we have a sequence (p_n), and we are interested in numbers that have a representation:

$$f = q_0 + \cfrac{p_1}{q_1 + \cfrac{p_2}{q_2 + \cfrac{p_3}{q_3 + \cfrac{p_4}{q_4 + \cfrac{p_5}{q_5 + \cdots}}}}} \qquad (9.2.2)$$

Unless stated otherwise, all continuous fractions from now on will be *regular*.

Theorem 9.2.1. Every positive rational number can be represented by a terminating continuous fraction.

Proof. Consider the rational number $f = \frac{a}{b}$ with $a > b$. We can apply Euclid's algorithm (as we did for $f = \frac{93}{36}$) to see that $f = [q_0, q_1, q_2, \ldots, q_N]$ with $q_N = $ hcf(a, b). On the other hand if $a < b$ we use

$$\frac{a}{b} = \frac{1}{\frac{b}{a}},$$

to see that if $\frac{b}{a} = [q_0, q_1, q_2, \ldots, q_N]$ then $\frac{a}{b} = [0, q_0, q_1, q_2, \ldots, q_N]$. $\square$

Theorem 9.2.2. There is no positive irrational number which can be represented by a terminating continuous fraction.

Proof. Let α be a positive irrational number. Then we can certainly write $q_0 < \alpha < q_0 + 1$, where q_0 is a nonnegative integer. Write $\alpha = q_0 + \rho_1$ where $0 < \rho_1 < 1$. So $\alpha = q_0 + \frac{1}{\beta_1}$ where $\beta_1 > 1$. Furthermore β_1 is irrational for if it is rational, then so is α and that's a contradiction. Write $\beta_1 = q_1 + \rho_2$ where q_1 is a nonnegative integer. We then have

$$\alpha = q_0 + \cfrac{1}{q_1 + \rho_2} = q_0 + \cfrac{1}{q_1 + \cfrac{1}{\beta_2}}.$$

Now β_2 is irrational and we can continue as above to get $\alpha = [q_0, q_1, q_2, q_3, q_4, \ldots]$. Now suppose this continued fraction terminates. Then $q_{N+1} = 0$ for some N and so $\beta_N = q_N$. But β_N is irrational and q_N is an integer and we have our desired contradiction. $\square$

In fact it can be shown that a positive irrational number can be written as the limit of the sequence of 'convergents' or to be precise:

$$\alpha = \lim_{N \to \infty} [q_0, q_1, q_2, \ldots, q_N].$$

This is the sense in which every positive irrational number can be represented as a 'non-terminating' continuous fraction. We will not give a proof of this fact here as the argument is quite lengthy. Instead we'll look at two delightful examples:

Example 9.1: The Square Root of Two

We know that $1 < \sqrt{2} < 2$ so we can write $\sqrt{2} = 1 + \frac{1}{\alpha_1}$, then

$$\alpha_1 = \frac{1}{\sqrt{2} - 1} = \frac{\sqrt{2} + 1}{(\sqrt{2} - 1)(\sqrt{2} + 1)} = \sqrt{2} + 1.$$

Since $2 < \sqrt{2} + 1 < 3$, we can write $\alpha_1 = 2 + \frac{1}{\alpha_2}$ so at this stage we have

$$\sqrt{2} = 1 + \cfrac{1}{2 + \cfrac{1}{\alpha_2}}.$$

But

$$\alpha_2 = \frac{1}{\alpha_1 - 2} = \frac{1}{\sqrt{2} - 1} = \sqrt{2} + 1.$$

So we can write $\alpha_2 = 2 + \frac{1}{\alpha_3}$ and the discerning reader will have noticed the emergence of a pattern. Indeed $\alpha_n = 2 + \frac{1}{\alpha_{n+1}}$ for all n and so it follows that

$$\sqrt{2} = [1, 2, 2, 2, \ldots, 2, \ldots],$$

i.e. $q_0 = 1$ and $q_n = 2$ for all $n \geq 1$.

Example 9.2: The Golden Section

We consider the golden section $\phi = \frac{\sqrt{5}+1}{2}$ and recall from Chapter 4, section 4 that $\frac{1}{\phi} = \frac{\sqrt{5}-1}{2}$. We know that $1 < \phi < 2$ and so we can write $\phi = \frac{\sqrt{5}}{2} + \frac{1}{2} = 1 + \frac{1}{\alpha_1}$. So $\frac{1}{\alpha_1} = \frac{\sqrt{5}}{2} - \frac{1}{2} = \frac{1}{\phi}$. Hence $\alpha_1 = \phi$ and so

$$\phi = 1 + \frac{1}{\phi} = 1 + \cfrac{1}{1 + \cfrac{1}{\phi}} = 1 + \cfrac{1}{1 + \cfrac{1}{1 + \cfrac{1}{\phi}}} = \cdots$$

Hence we see that

$$\phi = [1, 1, 1, 1, \ldots, 1, \ldots],$$

i.e. $q_n = 1$ for all n. Isn't that beautiful!

Both the examples of continued fractions that we've just looked at have a predictable pattern (at least after q_0 in the first case). We say that a continued fraction expansion is *periodic* if there exists N such that all numbers after q_N are just repetitions of the list $q_{N+1}, q_{N+2}, \ldots, q_{N+k}$ for some natural number k. For example you can check by using the methods employed above that $\sqrt{3} = [1, 1, 2, 1, 2, 1, 2, 1, 2, \ldots, 1, 2]$, so in this case $N = 0, k = 2$ and the repeating pattern is $1, 2$. The French mathematician Joseph-Louis Lagrange[1] (1736–1813) showed that a continued fraction is periodic if and only if it has the form $\frac{a+\sqrt{b}}{c}$ where a and c are integers and b is a natural number that is not a perfect square.

When we come to irrational numbers that are not given in terms of square roots, we cannot expect periodic behaviour and indeed the following results are known and are stated here without proof:

$$\pi = [3, 7, 15, 1, 292, 1, 1, 1, 2, 1, 3, \ldots]$$

$$e = [2, 1, 2, 1, 1, 4, 1, 1, 6, 1, 1, 8, 1, 1, 10, \ldots]$$

$$\gamma = [0, 1, 1, 2, 1, 2, 1, 4, 3, 13, 5, 1, 1, \ldots]$$

Of these three, the one for e has the nicest pattern. Bear in mind that the expansion for γ will terminate if it turns out that this number is rational.

---

[1] See e.g. http://en.wikipedia.org/wiki/Joseph_Louis_Lagrange

We can also obtain some very pleasant representations for π and e by using irregular continued fractions:

$$e = 2 + \cfrac{1}{1 + \cfrac{1}{2 + \cfrac{2}{3 + \cfrac{3}{4 + \cfrac{4}{5 + \cdots}}}}}$$

$$\frac{4}{\pi} = 1 + \cfrac{1}{2 + \cfrac{3^2}{2 + \cfrac{5^2}{2 + \cfrac{7^2}{2 + \cfrac{9^2}{2 + \cdots}}}}}$$

10

How Infinite Can You Get?

The infinite has always stirred the emotions of mankind more deeply than any other question; the infinite has stimulated and fertilized reason as few other ideas have; but also the infinite, more than any other notion, is in need of clarification.

On the infinite, David Hilbert[1]

It's time to return to the problem of infinity. In Chapter 2 we made a naive but unsatisfactory attempt to define infinity as $\frac{1}{0}$. The modern theory of the infinite dates back to groundbreaking work by Georg Cantor (1845–1918) during 1874–84.[2] He began by trying to figure out what we really mean by *counting*. Suppose that we have a supply of letters and let's write down six of them: *b, g, k, n, p, w*. How do we know that we really have six symbols? Of course we count them, but what does this really mean? Cantor proposed that the essence of counting resides in a one-to-one correspondence between natural numbers and the objects we are trying to count. So when we count the symbols we are doing something like this:

$$b \longleftrightarrow 1$$

$$g \longleftrightarrow 2$$

$$k \longleftrightarrow 3$$

$$n \longleftrightarrow 4$$

$$p \longleftrightarrow 5$$

$$w \longleftrightarrow 6$$

[1] This famous essay by one of the great mathematical minds of the early twentieth century, which was first published in 1925, is reprinted in *From Frege to Gödel: A Sourcebook in Mathematical Logic* ed. J. van Heijenoort, Harvard University Press (1967).

[2] See http://en.wikipedia.org/wiki/Georg_Cantor

In this procedure each symbol is mapped to a number between 1 and 6 and no symbol is mapped to more than one number. This is the essence of counting. We would implement the same process if we were counting cows in a field or pencils on my desk. In every case where we count a collection of discrete objects we put them in one-to-one correspondence with[3] the natural numbers starting at one. The last number we need, after which our collection of objects is exhausted, is its size.

I'll repeat a key point. The essence of a one-to-one correspondence is that every object is related by the arrow to one and only one number. If (for example) we used two colours, red and blue, to paint our symbols then we would not have a one-to-one correspondence between objects and colours as b, k, n and w might all be red while g and p are blue.

When counting is viewed as a one-to-one correspondence, Cantor used the word *cardinality* to describe it. So the number 6 is the cardinality of the collection b, g, k, n, p, w. If I take a smaller selection of these numbers e.g. g, p, w then it has a smaller cardinality – which is 3 in this case.

As long as we stick with finite collections,[4] then we cannot ever set up a one-to-one correspondence between a collection of objects and a sub-collection comprising some (but not all) of the original collection. When we extend to collections of infinite size, one of Cantor's greatest achievements was to turn this on its head. Let's start with the natural numbers $1, 2, 3, \ldots$ This is an infinite collection – indeed we saw way back in Chapter 1 that there is no largest element. Cantor defined the cardinality of the collection of all natural numbers to be $\aleph_0$. Here $\aleph$ (pronounced 'aleph') is the first letter in the Hebrew alphabet and the subscript 0 should just be accepted for now. Now at the moment $\aleph_0$ is just a symbol and you may be wondering why Cantor doesn't just write ∞?

Be aware that Cantor is using $\aleph_0$ as a symbol for infinity in a very special way. Let's go back to the number 3. From Cantor's point of view this is the property that all collections of three objects in the universe share in common. They all partake in 'threeness'. Similarly $\aleph_0$ should be the property that certain (why not all?) collections of infinite objects share in common. So any collection that can be put into one-to-one correspondence with the natural numbers will also have cardinality $\aleph_0$. Now consider the even numbers $2, 4, 6, 8, \cdots$ There are surely 'fewer of these' than there are of the natural numbers as we've omitted all the odd numbers, however they also have cardinality $\aleph_0$ since we have the one-to-one correspondence:

$$1 \longleftrightarrow 2, 2 \longleftrightarrow 4, 3 \longleftrightarrow 6, 4 \longleftrightarrow 8 \ldots,$$

indeed any natural number is mapped uniquely to an even number by the formula $n \longleftrightarrow 2n$. This example demonstrates clearly that when we get to infinite

---

[3] Technically – a subset of – see next footnote.
[4] Mathematicians use the technical term 'set' when they want to discuss collections of (mathematical) objects. See Appendix 2.

collections it is no longer true that the whole is greater than all of its parts. Indeed we've just shown that $\aleph_0 = \frac{\aleph_0}{2}$ and this should be compared with the 'algebra of infinity' that we discussed rather naively at the end of Chapter 2.

Now let's consider the integers. There should be $2\aleph_0 + 1$ of these as each natural number has a negative partner and we also want to include zero. From what you've just seen you might guess that $2\aleph_0 + 1 = \aleph_0$ and that is indeed the case as we have a one-to-one correspondence between natural numbers and integers given as follows:

$$1 \longleftrightarrow 0, 2 \longleftrightarrow 1, 3 \longleftrightarrow -1, 4 \longleftrightarrow 2, 5 \longleftrightarrow -2, 6 \longleftrightarrow 3, 7 \longleftrightarrow -3, \dots$$

If you want to express this by a concise formula, we have $n \longleftrightarrow \frac{n}{2}$ if n is even and $n \longleftrightarrow -\frac{n-1}{2}$ if n is odd. Here natural numbers are on the left-hand side of the arrow and integers on the right-hand side.

Now what about the rational numbers? Surely there are going to be more of these than the natural numbers. Cantor proved there wasn't. To see why let's first of all restrict to the positive rational numbers. We can write these in a list as follows:

$\frac{1}{1}$	$\frac{2}{1}$	$\frac{3}{1}$	$\frac{4}{1}$	$\frac{5}{1}$	$\frac{6}{1}$	$\frac{7}{1} \cdots$
$\frac{1}{2}$	$\frac{3}{2}$	$\frac{5}{2}$	$\frac{7}{2}$	$\frac{9}{2}$	$\frac{11}{2}$	$\frac{13}{2} \cdots$
$\frac{1}{3}$	$\frac{2}{3}$	$\frac{4}{3}$	$\frac{5}{3}$	$\frac{7}{3}$	$\frac{8}{3}$	$\frac{10}{3} \cdots$
$\frac{1}{4}$	$\frac{3}{4}$	$\frac{5}{4}$	$\frac{7}{4}$	$\frac{9}{4}$	$\frac{11}{4}$	$\frac{13}{4} \cdots$
$\cdots$	$\cdots$	$\cdots$	$\cdots$	$\cdots$	$\cdots$	$\cdots$
$\cdots$	$\cdots$	$\cdots$	$\cdots$	$\cdots$	$\cdots$	$\cdots$
$\cdots$	$\cdots$	$\cdots$	$\cdots$	$\cdots$	$\cdots$	$\cdots$

Each number is written in the form $\frac{p}{q}$. In the first row we have listed all the numbers for which $q = 1$ and these are of course the natural numbers. In the second row we have listed all the numbers for which $q = 2$ and which aren't already in row 1 – so we omit e.g. $\frac{4}{2}$ as it is already included as the number $\frac{2}{1}$. Clearly there are an infinite number of rows and each row is infinite. So the cardinality of the positive rational numbers is surely $\aleph_0^2 = \aleph_0 \times \aleph_0$. We now show that that is precisely $\aleph_0$. To see this we simply count the numbers in the order of increasing $p + q$ but where these are the same we count the number with the highest p first. So e.g. $\frac{4}{1}$ and $\frac{3}{2}$ both have $p + q = 5$ but since $4 > 3$ we count $\frac{4}{1}$ first. We then get the following one-to-one correspondence:

$$1 \longleftrightarrow \frac{1}{1}, \ 2 \longleftrightarrow \frac{2}{1}, \ 3 \longleftrightarrow \frac{1}{2}, \ 4 \longleftrightarrow \frac{3}{1}, \ 5 \longleftrightarrow \frac{1}{3}, \ 6 \longleftrightarrow \frac{4}{1},$$

$$7 \longleftrightarrow \frac{3}{2}, \ 8 \longleftrightarrow \frac{2}{3}, \ 9 \longleftrightarrow \frac{1}{4}, \ 10 \longleftrightarrow \frac{5}{1} \dots$$

You can include negative rational numbers and zero in this scheme by imitating the argument that we used above to count the integers.[5]

Ingenious as this argument is we seem to be discovering a predictable monotony about the infinite. Natural numbers, even numbers, integers and rational numbers all have the same cardinality. We say they are *countable* as they can all be put into a one-to-one correspondence with the natural numbers. But there is a twist in the tale. Consider the open interval $(0, 1)$. It *cannot* be put into a one-to-one correspondence with the natural numbers and Cantor called it an *uncountable set* for this reason. How do you prove $(0, 1)$ is uncountable? Write each point in the interval as an infinite decimal $0.x_1x_2x_3 \cdots$ Now suppose that $(0, 1)$ is countable. Then we have a one-to-one correspondence

$$
\begin{array}{rcl}
1 & \longleftrightarrow & 0.a_1a_2a_3a_4a_5 \cdots \\
2 & \longleftrightarrow & 0.b_1b_2b_3b_4b_5 \cdots \\
3 & \longleftrightarrow & 0.c_1c_2c_3c_4c_5 \cdots \\
4 & \longleftrightarrow & 0.d_1d_2d_3d_4d_5 \cdots \\
5 & \longleftrightarrow & 0.e_1e_2e_3e_4e_5 \cdots \\
\cdots & \cdots & \cdots \cdots \cdots \cdots \cdots \\
\cdots & \cdots & \cdots \cdots \cdots \cdots \cdots \\
\cdots & \cdots & \cdots \cdots \cdots \cdots \cdots \\
\end{array}
$$

It doesn't matter what specific numbers the as, bs, cs, ds and es represent. They are just numbers between 0 and 9. Now if $(0, 1)$ really is countable then every real number between 0 and 1 occurs on the right-hand side in this correspondence. Cantor used his famous *diagonal argument* to construct a number between 0 and 1 which is not on the list. His candidate was the number $0.x_1x_2x_3x_4x_5 \cdots$ where $x_1 \neq a_1, x_2 \neq b_2, x_3 \neq c_3, x_4 \neq d_4, x_5 \neq e_5$ etc. Now which number on the list is this one? As, $x_1 \neq a_1$ it can't be the first one, as $x_2 \neq b_2$, it isn't the second since $x_3 \neq c_3$, it won't be the third and so it goes. As the list is assumed to be complete we've deduced a contradiction.

Cantor assigned the letter c to be the cardinality of the interval $(0, 1)$. The letter c stands for 'continuum' and what we've just seen tells us that our notion of the infinite changes when we pass from the discrete to the continuous. $\aleph_0$ is the infinity of the discrete and c is the infinity of the continuous. In fact c is also the cardinality of the whole real line because it can be out into one-to-one correspondence with $(0, 1)$ through the formula $x \to \tan\left(\pi x - \frac{\pi}{2}\right)$. Since the real line contains the natural numbers as a subcollection we can assert that

$$c > \aleph_0.$$

In fact it can be shown that $c = 2^{\aleph_0}$ and this beautiful formula is embossed on Cantor's memorial in his home town of Halle. For a quick insight as to why this is true – write each non-zero real number between 0 and 1 in the form

[5] The scheme presented here is not the only way of enumerating the positive rational numbers and you may well meet others when you read different textbooks.

$0.a_1 a_2 a_3 \cdots$ where instead of the usual decimal expansion we use binary. So each a_n is either a 0 or a 1 and we have a one-to-one correspondence between the interval $(0, 1)$ and the collection of all binary sequences (a_n). Now there are two ways of choosing a_1, two ways of choosing a_2 etc. As there are $\aleph_0$ numbers in each sequence the cardinality of the collection of all of these is precisely $2^{\aleph_0}$ and so this is the cardinality of $(0, 1)$. From what has been written above we see that it is also the cardinality of the real line. Incidentally, if you remove all the rational numbers from the real line and just let the irrational numbers remain then the cardinality of these is also c and so

$$c - \aleph_0 = c.$$

So far our notion of counting and its extension to infinite collections has been based on the notion of one-to-one correspondence. It doesn't matter how we order the numbers 1, 2 and 3 – if it is 1, 2, 3 or 3, 1, 2 the cardinality of the collection is always 3. But this ignores the fact that we count numbers in order $1, 2, 3, \ldots$ and $1 < 2 < 3 < \cdots$ Cantor argued that counting numbers in order is a different process from matching them up in a one-to-one correspondence. When collections of numbers are finite, we don't notice this but when they are infinite the distinction matters. So imagine that we try to count the natural numbers *in order*. Then we get $1, 2, 3, 4, \ldots, 100, \ldots, 1000, \ldots, 1000000, \ldots, \omega_0$. Now ω_0 is an infinite number that is the end-product of counting all the natural numbers in order starting at the number 1. It is not the same as $\aleph_0$ which is the cardinality of a collection that is put into one-to-one correspondence with the natural numbers. Cantor argued that 'putting into one-to-one correspondence' and 'counting in order' are logically distinct processes and each has its own notion of 'infinity'. Cantor called ω_0 an *ordinal number*.

When we deal with cardinal numbers, the 'algebra of infinity' ensures that $\aleph_0 + 1 = 1 + \aleph_0$. But this is not the case with ordinal numbers. Once we have counted to ω_0, we can start again and go to the next number $\omega_0 + 1$ and then $\omega_0 + 2$ followed by $\omega_0 + 3$ etc. The cardinality of the collection of numbers (or 'set' if we are going to use the precise language of mathematicians) $1, 2, 3, \ldots, \omega_0, \omega_0 + 1, \omega_0 + 2, \omega_0 + 3$ is still $\aleph_0$ but the highest number we have reached in our counting is $\omega_0 + 3$ which is bigger than $\omega_0 + 2$ which is bigger than $\omega_0 + 1$ which is itself bigger than ω_0, where 'bigger' is understood solely in the sense of ordinal numbers. Furthermore $1 + \omega_0 \neq \omega_0 + 1$ as $1 + \omega_0$ is the end-point that we get to when we start counting in order from 2 as the starting point, and this is precisely ω_0. So $1 + \omega_0 = \omega_0$ which is smaller than $\omega_0 + 1$.

Now Cantor was able to show that as we carry on counting these 'transfinite ordinals' eventually we will reach an ordinal number called ω_1 which is such that the cardinality of the collection $1, 2, 3, \ldots, \omega_0, \ldots, \omega_1$ is no longer $\aleph_0$. He used the notation $\aleph_1$ to denote the cardinality of this set. Beyond ω_1 lies ω_2 and this leads to a collection with cardinality $\aleph_2$ and so we see the beginnings of the development of an infinite sequence of both transfinite ordinals and associated transfinite cardinals.

Where does c fit into this story? Cantor's famous *continuum hypothesis* conjectures that $c = \aleph_1$, but he was unable to prove this. In 1963, the logician Paul Cohen (1934–2007) showed that this conjecture was *undecidable* in that there are valid axiom systems for the foundations of mathematics under which it is true and others under which it is false. This doesn't bother most working mathematicians for whom the fine properties of higher orders of infinity are an irrelevance. But the fact that $\aleph_0$ and c are different is of great importance in modern mathematics.

In his famous essay 'On the infinite' which starts this chapter, David Hilbert makes the distinction between 'potential infinity' and 'actual infinity'.[6] Potential infinity has been the theme of much of this book. Mathematical analysis is based on the notion of the limit which avoids actual infinity by using only finite processes in the mathematical development. However the development of set theory at the end of the nineteenth century forced mathematicians to directly use collections of infinite objects in their reasoning and Cantor was then able to give 'actual infinity' a precise meaning. Not all mathematicians agreed with this but we will not go into the details of that saga here.[7] Hilbert wrote 'No-one shall expel us from the Paradise that Cantor has created' and contemporary mathematicians (as I pointed out above) are content to let this 'Paradise' be, though few venture very far into its garden of delights.

---

[6] In fact this distinction goes back to the Greek thinker Aristotle (384–22 BCE), who in the fourth century BCE wrote in his book *Physics*: 'The infinite has a potential existence ...there will not be an actual infinite'.

[7] Leonard Kronecker, who we met at the beginning of Chapter 2, was a strong opponent of Cantor's ideas.

11

Constructing the Real Numbers

Mathematical analysis is as extensive as nature herself.

Joseph Fourier

A key theme of this book has been the subtlety and beauty of the real number system. But what are real numbers anyway? We have seen that the rational numbers are inadequate for many of the purposes of mathematics. Indeed they cannot fill up the real number line. Irrational numbers can be obtained as limits of sequences of rational numbers using decimal expansions (or continued fractions). But this isn't really satisfactory as the notion of limit was itself based on the assumption that real numbers exist. The real numbers can be systematically built from the rational numbers using either of two approaches. One of these is the idea of a *Dedekind cut* which was introduced by Richard Dedekind (1831–1916)[1] and the other is by using *Cauchy sequences* and was due to Georg Cantor (1848–1918).[2]

We'll briefly sketch each of these. However I stress that this chapter is highly incomplete and gives only the skimpiest of introductions to a highly complex subject.

11.1 Dedekind Cuts

Let's mark all the rational numbers on the real number line. Of course this leaves holes where the irrational numbers need to go. Dedekind's idea is to use all the rational numbers that are smaller than the irrational number that we want to

---

[1] See http://en.wikipedia.org/wiki/Richard_Dedekind
[2] See http://en.wikipedia.org/wiki/Georg_Cantor

construct as a tool to fill the hole. To be more precise – we define a *cut c* to be a collection (or set) of rational numbers that has the following four properties:

(Ci) c contains at least one rational number.

(Cii) If p is in c then so is any rational number smaller than p.

(Ciii) c has no largest element, so if p is in c then there exists a rational number q such that $q > p$ and q is in c.

(Civ) c does not contain all the rational numbers.

We now define the real number line to be the collection of all possible cuts. This is a radical re-interpretation of number. We might ask what $\sqrt{2}$ is. The answer is that it is the cut that contains all rational numbers p for which $p^2 < 2$. So for example 1 is in this cut, so is 1.4 and so is 1.4142136. All the rational numbers themselves are naturally associated with cuts, so the rational number q is identified with the cut containing all rational numbers p such that $p < q$, so for example the cut which represents 1.5 contains the numbers 1.49, 1.499 and 1.499999999999999, but it does not contain 1.5 itself.

A natural question we might ask is if cuts are really going to represent real numbers then how do we do arithmetic with them? Well we have to redefine addition and multiplication, so for example if c and d are cuts then $c + d$ is the cut which contains all rational numbers $p + q$ where p is in c and q is in d. Multiplication is a little more complicated to define. Nonetheless it can be shown that these definitions coincide with our usual notions (so e.g. $2 + 3$ still equals 5) and all the usual algebraic laws of numbers continue to hold for cuts such as $a(b + c) = ab + ac$. Also if c and d are cuts we say that $c < d$ if d contains at least one rational number that is larger than every rational number in c.

It's important to appreciate that the purpose of cuts is to give us a systematic approach to defining real numbers from rationals. They will not help us to do practical calculations with numbers but they do give us a firm foundation on which to rest the vast edifice of theorems about real numbers, including those we've met earlier in this book.

11.2 Cauchy Sequences

An alternative way of constructing the real numbers other than using cuts is to employ Cauchy sequences. Let's first define this concept within the spirit of Chapter 4 where we assume that real numbers exist. Then we'll see how the concept can be tweaked to actually produce real numbers from rationals. We say that the sequence (x_n) is a *Cauchy sequence* if given any $\epsilon > 0$ there exists a natural number N such that $|x_n - x_m| < \epsilon$ whenever both of $m, n > N$. This looks remarkably like the definition of a limit except that we haven't asked that the sequence converges. A Cauchy sequence has the property that if you go far enough along it, then the terms of the sequence become arbitrarily close. It is easy to prove that every convergent sequence is Cauchy. For suppose that (x_n) converges to a limit l then using the MFT and the triangle inequality

$$|x_n - x_m| = |(x_n - l) + (l - x_m)| \leq |x_n - l| + |l - x_m|,$$

and I'll leave the rest to you. Conversely, Augustus-Louis Cauchy (1789–1857) proved that every Cauchy sequence converges to a limit but he was assuming that real numbers exist. Let's drop that assumption and return to a world where the only numbers are the rationals.

Cantor first redefined Cauchy sequences using rational numbers only. So for the rest of this section a Cauchy sequence is a sequence (x_n) of rational numbers which has the property that, given any rational $\epsilon > 0$, there exists a natural number N such that $|x_n - x_m| < \epsilon$ whenever both of $m, n > N$. Cantor's idea was to define the real number line as the collection of all (rational) Cauchy sequences. So for example the rational number 1 should be identified with the Cauchy sequence $(1, 1, 1, \ldots)$. But there is a problem here, as the geometric series $\sum_{n=1}^{\infty} \frac{1}{2^n}$ also converges to 1. To overcome this problem we need one more definition. Two Cauchy sequences (x_n) and (y_n) are said to be *equivalent* if, given any rational $\epsilon > 0$, there exists a natural number N such that if $n > N$ then $|x_n - y_n| < \epsilon$. This is clearly the case for the two sequences above which both converge to 1. Cantor's approach was to identify equivalent Cauchy sequences as representing the same real number.[3] So $\sqrt{2}$ is identified with the Cauchy sequence that begins 1, 1.4, 1.41, 1.414, 1.4142, 1.41421, 1.414213, 1.4142136, $\ldots$) and also with the sequence (x_n) defined recursively by $x_1 = 1$ and for $n > 1$, $x_{n+1} = \frac{x_n}{2} + \frac{1}{x_n}$ (see Section 5.4). If we use Cauchy sequences to define real numbers then addition and multiplication are both fairly easy – so if the Cauchy sequences (a_n) and (b_n) represent the real numbers a and b respectively then $a + b$ is represented by the Cauchy sequence whose nth term is $a_n + b_n$, while ab corresponds to the Cauchy sequence with nth term $a_n b_n$. To capture the notion of 'less than' we need to be more subtle – we say that $a < b$ if given any rational $\epsilon > 0$ there exists a natural number N such that if $n > N$ then $b_n - a_n > \epsilon$.

11.3 Completeness

Both definitions of real numbers that we have given require us to think about numbers in radically new ways. They each employ sets of rational numbers to

---

[3] This really requires the notion of an *equivalence relation*.

represent real numbers – either gathered together in cuts or assembled into Cauchy sequences.[4] However you define real numbers, the power of the definition rests in its ability to give a thorough logical foundation for further investigation. One of the most important results, which flows from the construction of the real numbers from either Dedekind cuts or Cauchy sequences, is the *completeness property* of the real numbers. We will now prove this important property using Dedekind cuts.[5]

Theorem 11.3.1. Let A be a non-empty collection (or set) of real numbers that is bounded above. Then it has a least upper bound.

Let's be clear what the theorem says. A non-empty collection of numbers means that we have assembled at least one (and possibly infinitely many) numbers and we call this collection A. Bounded above means that there exists a real number K such that if a is any of the numbers that is in A then we must have $a < K$. Finally to have a least upper bound means that the collection of all possible K that do this job has a smallest member which we call $\sup(A)$. A sequence (a_n) is an example of a collection of numbers so Theorem 11.3.1 guarantees that every sequence that is bounded above has a supremum and this underlines everything we did in Chapter 5.

Proof of Theorem 11.3.1. In this proof I'll have to use set theoretic notation (see Appendix 2). So if A is a set of numbers then $B \subset A$ means B is a (proper) subset of A (i.e. every number in B is also in A but A contains at least one number that is not in B) and $A \subseteq B$ means either $A \subset B$ or $A = B$.

Now let A be a non-empty subset of the real numbers so A contains at least one cut. Assume that A is bounded above and let β be an upper bound for A. This means that $c \subseteq \beta$ for every cut c in A. Define α to be the set that contains every rational number in every cut in A (i.e. α is the union of all the cuts in A). We will show firstly that α is a cut and secondly that $\alpha = \sup(A)$. Then the theorem will be proved.

To show that α is a cut we must check that (Ci) to (Civ) are satisfied. Let c be a cut in A. Every rational number q in c is also in α so (Ci) is satisfied. If q is any rational number in α then q lies in one of the cuts that form α, and so every rational number smaller than q is in that cut and hence in α, so (Cii) is satisfied. For (Ciii) if α contains a largest rational number then so does one of the cuts that form α and that's a contradiction. Finally, it is clear that $\alpha \subseteq \beta$ and (Civ) follows as any rational number that is larger than every rational number in β is certainly not in α.

---

[4] Technically speaking – equivalence classes of these.
[5] You can find an online proof using Cauchy sequences sketched on http://en.wikipedia.org/wiki/Construction_of_the_real_numbers#Construction_from_Cauchy_sequences

Now to show $\alpha = \sup(A)$, suppose that $y < \alpha$ (as numbers). We'll show that y cannot be an upper bound for A. As $y \subset \alpha$ (as sets) there must exist a cut c' in α so that c' contains a rational number that is greater than every rational number in y. But then y cannot possibly be an upper bound for A and so we are done. $\quad\square$

Once we have Theorem 11.3.1 we can see that a Dedekind cut c is essentially identified with the unique real number $\sup(c)$. Although we won't prove it here, it's an important fact that the completeness property is equivalent to the statement that every Cauchy sequence has a limit in the real numbers.

One of the consequences of completeness (Theorem 11.3.1) is a key property of the real numbers that is named in honour of the great Archimedes of Syracuse (c.287–c.212 BCE). We used this property without proof in Chapter 4 to show that $\lim_{n \to \infty} \frac{1}{n} = 0$. Now let's see how it is derived from Theorem 11.3.1.

Theorem 11.3.2 (Archimedean Property of Real Numbers). Let x and y be arbitrary positive real numbers. Then there exists a natural number n such that $nx > y$.

Proof. Suppose that the statement is false so that $nx \le y$ for all n. Then the set A of all numbers of the form nx (where n is a natural number) is non-empty (it contains x) and is bounded above by y. Hence A has a least upper bound by Theorem 11.3.1. Write $\alpha = \sup(A)$ and pick any natural number n. Then $(n+1)x$ is in the set A and so $(n+1)x \le \alpha$. It follows that

$$nx \le \alpha - x < \alpha.$$

As this argument works for any n we see that $\alpha - x$ is also an upper bound for A and that contradicts the fact that α is the smallest of these. $\quad\square$

We'll finish this chapter with a delightful and intriguing property of the real numbers. This is sometimes referred to as the *density* of the rational numbers in the real numbers.

Theorem 11.3.3. Given any two real numbers x and y with $y > x$ there exists a rational number q such that

$$x \le q \le y$$

Proof. As $y - x > 0$ we can apply Theorem 11.3.2 to find a natural number n such that $n(y - x) > 1$, i.e.

$$nx + 1 < ny \ldots \quad \text{(i)}$$

If $x > 0$ we can use Theorem 11.3.2 again to show that there exists a natural number m_1 such that $m_1 > nx$, and then any natural number m_2 has the property that $m_2 > -nx$. A moment's reflection convinces us that we can drop the requirement $x > 0$ and two such numbers still exist. So $-m_2 < nx < m_1$ for

155

any real number x. Since nx lies between two integers, we can narrow this down and find an integer m such that $m - 1 \leq nx \leq m$, i.e.

$$m \leq nx + 1 \leq m + 1 \qquad \text{(ii)}$$

Combining (i) and (ii) together we get $nx \leq m < ny$, and so

$$x \leq \frac{m}{n} < y.$$

Hence $\frac{m}{n}$ is our required rational number. □

If x and y are both irrational numbers then Theorem 11.3.3 tells us that there exists a rational number q such that $x < q < y$, i.e. there is a rational number between every pair of irrationals. On the other hand, in Theorem 3.1.1 we showed that there are infinitely many irrational numbers between every pair of rationals. This seems to suggest that there are as many rationals as irrationals, but we saw in Chapter 9 that the irrationals are of a higher order of infinity than the rationals. This gives us yet more insight into how complex and counter-intuitive the structure of the real number system is.

12

Where to Next in Analysis? The Calculus

Next to the creation of Euclidean geometry, the calculus has proved to be the most original and fruitful concept in all of mathematics.

Mathematics and the Physical World, M. Kline

The end of the last chapter was the effective conclusion of this book. But some readers may be eager for more and this chapter is an attempt to answer the question – where do I go from here? The simple answer is learn calculus (if you haven't already done so), pick up a more advanced textbook on analysis (see Further Reading for some suggestions) and start to work systematically through it. In this short chapter I'll just give a sneak preview of some of the key new ideas that you'll meet next.

12.1 Functions

The function concept has been something of a 'ghost at the feast' so far and yet it is one of the most fundamental ideas in the whole of mathematics. Functions[1] are the way in which mathematics describe relationships. In the simplest form these will be relationships between two variables x and y. As we explore different values of x, the variable y also changes in a predictable (but perhaps quite complicated) manner. The way in which the value of y depends on that of x is expressed through the function concept and we write $y = f(x)$ to formally describe the relationship. This is shorthand for 'y is a function of x'. To give a logically watertight definition of a function needs more set theory than I want to go into here so we'll skip

---

[1] Nowadays these are often called 'mappings'.

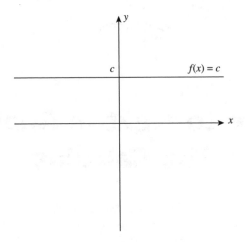

Figure 12.1. $f(x) = c$ with $c > 0$.

that and concentrate on examples.[2] To see the simplest function just pick a real number c and define $f(x) = c$. This is called the 'constant function', as each real number x is inexorably mapped to the number c. We get insight into functions by drawing their graphs and the constant function is shown in Figure 12.1.

Our next example, shown in Figure 12.2, is $y = mx$ where m is fixed. If (say) $m = 2$ we have $y = 2x$ and so $f(0) = 0, f(1) = 2, f(-0.48) = -0.96, f(2) = 4$ etc. The graph of the function is a straight line through the origin which has positive slope if $m > 0$ and negative slope if $m < 0$.

At this stage, it is worth pointing out that $f(x)$ and f are different mathematical entities which should not be confused with each other. $f(x)$ is a number – it is the *value* of the function at the point x. On the other hand, f is our symbol for the function itself which represents a relationship. It is not a number and can be seen as a higher-order mathematical object. Later on we will look at the next stage where we will meet an *operator*. Just as functions take numbers to numbers so (at the next level up), operators map functions to functions.

Now let's get back to examples. We've met $f(x) = c$ and $f(x) = mx$. Let's combine them and consider $f(x) = mx + c$. This is the general *linear* function, shown in Figure 12.3. Any infinite straight line drawn in the plane appears as the graph of such a function. As above it has slope m, and c measures the distance from the origin to the point where the line meets the y axis.

If the highest power of x that appears in the definition of a function is 2 then the function is said to be *quadratic*. The simplest such function is $f(x) = x^2$ and the most general is $f(x) = ax^2 + bx + c$ where a, b and c are real numbers with $a \neq 0$. In this case the graph (Figure 12.4) is always a parabola.

---

[2] In general, we can make sense of the function concept as a mapping between two arbitrary sets.

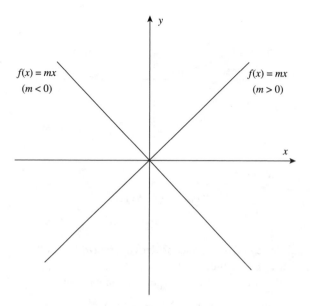

Figure 12.2. $f(x) = mx$.

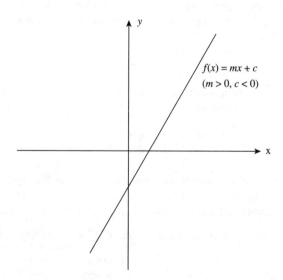

Figure 12.3. $f(x) = mx + c$.

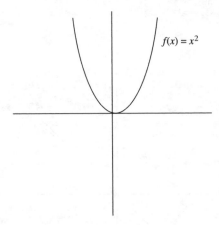

Figure 12.4. $f(x) = x^2$.

At this stage we could go on to consider general *cubic* functions, $f(x) = ax^3 + bx^2 + cx + d$ or *quartic* functions $f(x) = ax^4 + bx^3 + cx^2 + dx + e$, but let's go even further and write down the most general *polynomial*

$$f(x) = a_n x^n + a_{n-1}x^{n-1} + \cdots + a_2 x^2 + a_1 x + a_0,$$

where $a_0, a_1, a_2, \ldots, a_{n-1}, a_n$ are real numbers. If $a_n \neq 0$ then this is called a polynomial of *degree n*, e.g. $f(x) = x^{24} - 17x^9 + 14.673$ is a polynomial of degree 24. We can write the general polynomial more succinctly using sigma notation:

$$f(x) = \sum_{r=0}^{n} a_r x^r,$$

and it is natural to ask if we can generalise even further and consider functions that are defined by infinite series:

$$f(x) = \sum_{r=0}^{\infty} a_r x^r.$$

We have to be careful here for we need the series on the right-hand side to converge for a range of values of x. Any function f which is legitimately defined by a properly convergent series is said to have a *power series expansion*.[3] In fact, the general theory of power series tells us that if $\sum_{r=0}^{\infty} a_r x^r$ converges then it does so for all x in an *interval of convergence* $(-R, R)$. The positive real number R is called the *radius of convergence*. It then follows that the series diverges if $|x| > R$. If $x = R$

---

[3] The terminology 'power series' comes from the fact that we are using a series that involves powers of x.

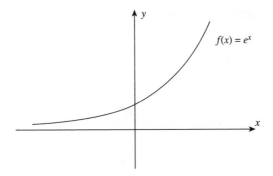

Figure 12.5. $f(x) = e^x$.

and $-R$ the general theory is inconclusive and the series may either converge or diverge, so a case-by-case consideration is needed. If the series converges for all values of x we say that $R = \infty$. An example of such a function that we've already met in Chapter 7 is the *exponential function* $f(x) = e^x$ (Figure 12.5). We recall that in this case:

$$e^x = \sum_{n=0}^{\infty} \frac{x^n}{n!}.$$

Two other well-known functions which converge for all values of x are the trigonometric functions $f(x) = \sin(x)$ and $f(x) = \cos(x)$, as shown in Figure 12.6. We first meet these as expressing important ratios of the sides of a right-angled triangle in basic trigonometry. When they are revealed as functions they are basic

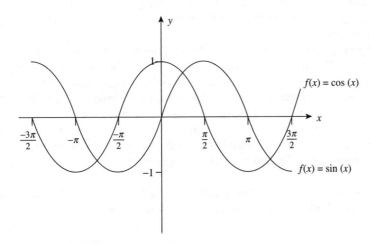

Figure 12.6. $f(x) = \sin(x)$ and $f(x) = \cos(x)$.

161

models for periodic phenomena such as wave motion. In this case the power series expansions are

$$\sin(x) = \sum_{n=0}^{\infty} (-1)^n \frac{x^{2n+1}}{(2n+1)!}, \qquad \cos(x) = \sum_{n=0}^{\infty} (-1)^n \frac{x^{2n}}{(2n)!}.$$

For an example of a function that has a radius of convergence $R < \infty$, consider $f(x) = \frac{1}{1-x}$. By the binomial series,[4] we have

$$\frac{1}{1-x} = \sum_{n=0}^{\infty} x^n,$$

for $-1 < x < 1$. If $|x| \geq 1$, you can see for yourself that the series diverges. So $R = 1$ in this case.

12.2 Limits and Continuity

Analysis really takes off when the concepts of function and limit meet and interact. We begin by defining the *limit of a function* at a point x. Let (x_n) be any sequence that converges to the point x. Now consider the sequence $(f(x_n))$. If $(f(x_n))$ converges to a real number l, for every sequence (x_n) that converges to x, we say that the function has a limit l at the point x, and we write

$$\lim_{y \to x} f(y) = l.$$

It turns out that this is equivalent to the following $\epsilon - \delta$ formulation: we say that $\lim_{y \to x} f(y) = l$ if given any $\epsilon > 0$, there exists $\delta > 0$ so that whenever $|y - x| < \delta$ we must have $|f(y) - l| < \epsilon$.

One of the most important applications of the limit concept is to the concept of *continuity*. The idea is to describe (using precise mathematics) a function that is continuous, i.e. it has the property that its graph can be drawn using a pencil (say) in one continuous flow without ever taking the pencil from the paper. We say that a function f is continuous if it is continuous at every point x and it is continuous at x if $\lim_{y \to x} f(y) = f(x)$. So the function f is continuous at x if for every sequence (x_n) that converges to x we must have that the sequence $(f(x_n))$ converges to $f(x)$. All polynomials are continuous and so are the functions $f(x) = e^x, f(x) = \sin(x)$ and $f(x) = \cos(x)$.

Here's an example of a function that fails to be continuous. In fact it doesn't even have a limit at $x = 0$. It is called the *Heaviside function*, after the engineer

---

[4] See Appendix 1 if you aren't familiar with this.

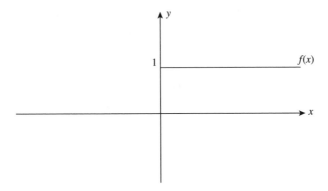

Figure 12.7. The Heaviside Function.

Oliver Heaviside (1850–1920) who first made use of it in applications.[5] It is defined by:

$$f(x) = \begin{cases} 0 & \text{if } x < 0 \\ 1 & \text{if } x \geq 0 \end{cases}$$

The graph of f is shown in Figure 12.7.

The Heaviside function f can represent a signal that is turned on to have a value of 1 unit at the point $x = 0$. From the graph it seems obvious that f is not continuous at $x = 0$. To see how to prove this rigorously, consider the sequence (x_n) where each $x_n = -\frac{1}{n}$. Clearly (x_n) converges to 0. Now for each n, $f(x_n) = f\left(-\frac{1}{n}\right) = 0$ and so we also have that $(f(x_n))$ converges to 0. But $f(0) = 1$ and so we have found a sequence (x_n) which converges to 0 but $(f(x_n))$ doesn't converge to $f(0)$. So f cannot be continuous at the point $x = 0$.[6]

12.3 Differentiation

In this section and the next, we'll briefly describe how the conceptual framework of analysis enables us to give rigorous meaning to the two key ideas of the calculus – differentiation and integration.

Differentiation is the process whereby we calculate the rate of change of a function at a point. The diagram in Figure 12.8 may give some insight into this.

---

[5] He also discovered the Heaviside layer in the atmosphere, see http://en.wikipedia.org/wiki/Oliver_Heaviside. Nowadays mathematicians often use the terminology *indicator function* to describe functions that take the value 1 on some interval and are zero for all other values of x.

[6] Of course f is continuous at every other point. Can you prove this using precise mathematics?

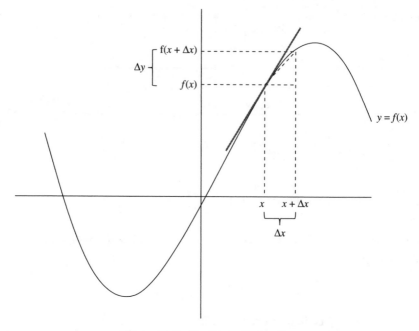

Figure 12.8. The slope of a function.

We consider two points on the x-axis which are a very small distance apart. It's traditional to use the symbol Δx to represent a small distance. In the diagram $\Delta x > 0$ and so our two points that are very close together are x and $x + \Delta x$. If $y = f(x)$, we'll define $\Delta y = f(x + \Delta x) - f(x)$. If f is a continuous function, then $\lim_{\Delta x \to 0} \Delta y = 0$. The rate of change of the function between the two points x and Δx is $\frac{\Delta y}{\Delta x}$. We want to measure the rate of change at the point x itself. Geometrically this is precisely the slope of the tangent line to the curve f at the point x. We might argue that this should be the value of $\frac{\Delta y}{\Delta x}$ at $\Delta x = 0$, but if f is continuous, this is $\frac{0}{0}$ which has no meaning. Isaac Newton (1643–1727) and Gottfried Leibniz (1646–1716) (working independently)[7] realised that[8] although $\frac{\Delta y}{\Delta x}$ has no meaning at $\Delta x = 0$, it is quite possible for $\lim_{\Delta x \to 0} \frac{\Delta y}{\Delta x}$ to exist and indeed this is the case for many important functions. To be precise, we say that a function f is *differentiable* at the point x if this limit exists and in this case we define

$$f'(x) = \lim_{h \to 0} \frac{f(x + h) - f(x)}{h}.$$

[7] I'm not going to comment on the priority dispute that so obsessed many of their followers and remains the subject of scholastic enquiry to this day.

[8] Of course, Newton and Leibniz did not know the precise definition of a limit but they had sufficient insight to be able to make the methodological breakthrough that established the calculus as a working tool for mathematicians and scientists.

If $f'(x)$ exists at every x then f' is a new function called the 'derived function' or *derivative* of f at the point x. To symbolise the fact that f' is a limit of ratios, we often use the convenient notation $\frac{dy}{dx} = f'(x)$; however I would emphasise that $\frac{dy}{dx}$ is a holistic notation. There is no mythical 'dy' that is to be divided by an equally mythical 'dx'. The table below gives some derivatives of well-known functions:

$f(x)$	$f'(x)$
c	0
x^n	nx^{n-1}
$\log(x)$	$\frac{1}{x}$
e^x	e^x
$\sin(x)$	$\cos(x)$
$\cos(x)$	$-\sin(x)$

As a fun exercise, take the fact that the derivative of x^n is nx^{n-1} as given and try to derive the derivatives of e^x, $\sin(x)$ and $\cos(x)$ by using the series expansions for these functions that were given in the previous section. If you think carefully about this you'll see that it involves interchanging two limiting procedures and that really needs to be justified carefully.

The importance of the derivative in science, engineering, economics, etc. arises from the fact that rates of change are so commonly encountered in applications. For example, if we take t to be time and y to be the displacement of a moving object from its starting point then $\frac{dy}{dt}$ is precisely the instantaneous velocity of the moving object. If we have an electric circuit and Q is the quantity of charge moving past at time t then $\frac{dQ}{dt}$ is the electric current.

Returning to mathematics, it can be shown that if a function f is differentiable at a point x then it must be continuous there, so (reversing the logic) discontinuous functions cannot have derivatives.[9] Differentiation is an example of an *operator* which is a function that changes functions to other functions. The study of general operators is a major part of modern analysis that forms a subject of its own which is called *functional analysis*. This is vital for a wide range of applications – especially *quantum theory*, the study of matter at the molecular, atomic and sub-atomic levels.

---

[9] At least not at the points of discontinuity.

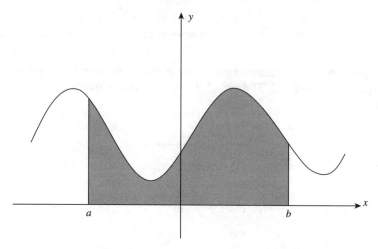

Figure 12.9. The area under a curve.

12.4 Integration

The roots of integration go back to the Greek geometer Eudoxus of Cnidus (c.408–c.355 BCE) (see Chapter 13) but we'll again pick up the story in the seventeeth century. The problem is to calculate the area under a curve which is described by a function f. For convenience we'll assume that $f(x) \geq 0$ for all x and that we want to calculate the area between the points $x = a$ and $x = b$, as shown in Figure 12.9.

Again the key breakthrough in methodology was made by Newton and Leibniz, but the solid foundations in analysis had to wait until the work of Riemann.[10] To see how to calculate the area under the curve we'll assume that the function f is *bounded* on the interval $[a, b]$ so there exist two nonnegative numbers m and M such that $m \leq f(x) \leq M$ for all $a \leq x \leq b$. We need to *partition* the interval $[a, b]$ into a set of points $\mathcal{P} = \{x_0, x_1, \ldots, x_n, x_{n+1}\}$, where we insist that $x_0 = a$, $x_{n+1} = b$ and $x_0 < x_1 < \cdots, < x_n < x_{n+1}$. We will not assume that the points in the partition are equally spaced. We define a crude approximation to the area under the curve (Figure 12.10) by overshooting on each interval $[x_j, x_{j+1}]$. To do this we define $M_j = \sup_{x_j \leq x \leq x_{j+1}} f(x)$. As f is bounded above, M_j is finite for all $j = 0, 1, \ldots, n$. We add up the area of all the rectangles that overshoot and this is precisely $\sum_{j=0}^{n} M_j(x_{j+1} - x_j)$.

---

[10] The approach that we'll give is in fact due to Jean-Gaston Darboux (1842–1917), see http://en.wikipedia.org/wiki/Jean_Gaston_Darboux

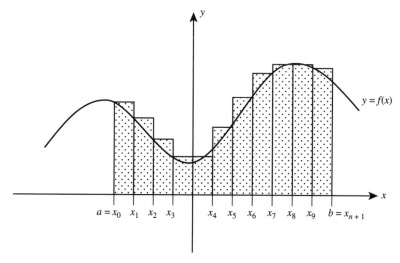

Figure 12.10. Overshooting the area.

Now imagine that we carry out this same calculation for all possible partitions of $[a, b]$. Clearly some overshoot approximations will be better than others in the sense that they get closer to the area that we want. We define the *upper integral* to be the best approximation that we can get in this way, i.e. the number which delicately undercuts all the overshoots. This is

$$\overline{\int_a^b} f(x)dx = \inf_{\mathcal{P}} \sum_{j=0}^{n} M_j(x_{j+1} - x_j).$$

The symbol on the left is the notation for upper integrals. Indeed we should think of an integral as a continuous version of a sum and Leibniz had the idea of representing this by an elongated S that is stretched out at both the top and bottom.[11]

Having got the best possible approximation to the area from above, let's do the same thing from below. Now we systematically undershoot and for a fixed partition $\mathcal{P}$ we define the undershoot area to be $\sum_{j=0}^{n} m_j(x_{j+1} - x_j)$ where $m_j = \inf_{x_j \le x \le x_{j+1}} f(x)$.

---

[11] The extension of the Leibniz notation to upper and lower integrals seems to have been introduced by Darboux.

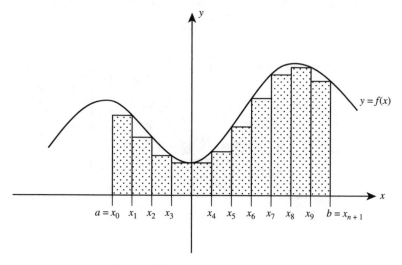

Figure 12.11. Undercutting the area.

The *lower integral* will delicately overtrim all the undershoots and this is defined by

$$\underline{\int_a^b} f(x)dx = \sup_{\mathcal{P}} \sum_{j=0}^{n} M_j(x_{j+1} - x_j).$$

Now it's a deep insight due to Darboux that the area under a curve makes sense (remember f is bounded but may not be continuous) only when delicately undercutting all the overshoots (see Figure 12.11) and just as delicately overtrimming all the undershoots gives precisely the same answer. So we say that a function f is *integrable* on $[a, b]$ if $\overline{\int_a^b} f(x)dx = \underline{\int_a^b} f(x)dx$. In this case we write $\int_a^b f(x)dx$ for the common value and call it the *integral* of the function f over the interval $[a, b]$. It is a fact that if f is continuous on such an interval then it is integrable.

We can see how the area grows by changing the fixed point b to a variable y. It turns out that there is always a continuous function F such that

$$\int_a^y f(x)dx = F(y) - F(a). \tag{12.4.1}$$

The function F is called an *indefinite integral* of f.[12]

Note from (12.4.1) that the function F is not unique – if F is a definite integral then so is $F + k$ where k is any fixed real number. The number k is usually

---

[12] It is also sometimes called a *primitive* or *antiderivative*.

called an *arbitrary constant*. In applications, it generally has a specific value that is determined by the way the world is. Some values of indefinite integrals for standard functions are shown below (without k).

$f(x)$	$F(x)$	
c	cx	
x^n	$\frac{x^{n+1}}{n+1}$	$(n \neq -1)$
$\frac{1}{x}$	$\log(x)$	
e^x	e^x	
$\sin(x)$	$-\cos(x)$	
$\cos(x)$	$\sin(x)$	

In basic calculus courses, the indefinite integral is often written with the same symbol as the definite one but without the upper and lower limits b and a:

$$\int f(x)dx = F(x) + k.$$

It is common to teach the method of going from f to F first as the reverse operation to differentiation and then to make the connection afterwards with the area under the curve $y = f(x)$ between $x = a$ and $x = b$ by arguing that $\int_a^b f(x)dx = F(b) - F(a)$. The work of Riemann, Darboux and their followers shows that from a foundational viewpoint, the definite integral is more fundamental and the indefinite integral derives its meaning from that more basic concept.

If you compare the two tables of derivatives and integrals you can indeed check that these are mutually inverse to each other, e.g. if we start with x^n ($n \neq 1$) and integrate we get $\frac{x^{n+1}}{n+1}$. Now differentiate this and we come back to x^n. This extends to an important general result. I pointed out earlier that if f is integrable, then F is continuous. Moreover if f is continuous it turns out that F is differentiable and

$$\frac{dF(x)}{dx} = f(x).$$

Conversely the so-called *fundamental theorem of calculus* tells us that if f is differentiable (hence continuous, and so integrable)

$$\int_a^b f'(x)dx = f(b) - f(a).$$

This is a good place to stop – but in truth, we have only just started. We haven't even mentioned some key highlights of calculus such as the mean value theorem and Taylor's theorem, then we should extend to two or more dimensions and look at partial derivatives and multiple integrals and also study calculus and analysis in the complex plane. All of these topics (and more) are vitally important for both pure and applied mathematics. The libraries are stacked with books, the internet is awash with references and the gentle reader is invited to go forth and sample the delights that await them.

13

Some Brief Remarks About the History of Analysis

The history of mathematics is itself a vast subject. To isolate just one topic – analysis – and try to outline its history is a somewhat dangerous enterprise and the reader should be aware that the author is not a professional historian and is a lot less sure of his footing in this territory. Analysis did not really emerge as an area of mathematics with its own identity until the nineteenth century. Although we can trace the development of some important ideas from antiquity it is difficult to disentangle these from geometry and algebra (and later – calculus). Nevertheless, if the reader holds these caveats firmly in his or her mind, we'll try to sketch our outline.

The main mathematical story in this book concluded with integration. Maybe the history of analysis should start there for what is integration but the art of calculating areas (and volumes) of complicated shapes by using limits of areas of simpler shapes? The first to have used this idea systematically seems to be the Greek geometer Eudoxus of Cnidus (c.408–c.355 BCE). He saw, for example, that you could get a good approximation to the area of a circle by filling it up with inscribed polygons with increasing numbers of sides. Books XI, XII and XIII of Euclid's celebrated *Elements* collects results of this type and the method was taken to even greater refinement by Archimedes of Syracuse (287–212 BCE). But I should emphasise that there was no formal limit concept in their work as has been developed in this book. The mathematical perspective in those days was geometric.

From Greek mathematics, we now fast forward to the rise of the modern world in the Western hemisphere. The fifteenth and sixteenth centuries saw a renewed interest in Greek geometry and significant progress in algebra. Rene Descartes (1596–1650) began the process of unifying these through his introduction of co-ordinates. The work of Galileo Galilei (1564–1642) in mechanics started to stimulate mathematicians to worry about how to calculate instantaneous velocities. The origins of calculus can be found in the work of Pierre de Fermat (1601–65), Descartes and Isaac Barrow (1630–77) who was Newton's predecessor at Cambridge, but their approach to the calculation of tangents to a curve was still dominated by geometric arguments. Important contributions were also made by John Wallis (1616–1703).

The development of a universal methodology that forged the calculus into a powerful working tool was due to Isaac Newton (1642–1727) and Gottfried Leibniz (1646–1716). The function concept was not yet established. Newton wrote of a changing quantity which he called a 'fluent' and its derivative was a 'fluxion' (in his work differentiation was always with respect to time). To differentiate the fluent f, Newton allows $f(x)$ to flow into $f(x + o)$ and he writes 'Let now the increment vanish ...' when he extracts the fluxion from the ratio $\frac{f(x+o)-f(x)}{o}$. This became known as the 'method of infinitesimals' and here lies the origin of the modern definition of the derivative as a limit. Leibniz introduced the notation $\frac{dy}{dx}$ for derivatives and $\int$ for integrals and was the first to see that differentiation and integration are mutually inverse.

Calculus was intensively developed by the followers of Newton and Leibniz but the mathematics was not rigorous and attracted some criticism from e.g. the philosopher and cleric George Berkeley (1685–1753) who famously argued that infinitesimals had the status of 'the ghost of departed quantities'. Towards the end of the eighteenth century it was clear that there was a notion of 'limit' that lay behind the unsatisfactory notion of infinitesimals, but it wasn't yet clear how to define it! In his article 'Limite'[1] Jean d'Alembert (1717–83) wrote 'The theory of limits is the true metaphysics of the calculus ...it is never a question of infinitesimal quantities in the differential calculus, it is uniquely a question of limits of finite quantities'.

Leonhard Euler (1707–83) was called 'analysis incarnate' by his contemporaries. He published an important two-volume treatise in 1748 called *Introduction to Infinitesimal Analysis*. This book pioneered the concept of a function as a relationship between two variables and also contained many results about infinite series and products and continued functions. Euler's work exhibited a wonderful genius for calculating explicit values of series and integrals (some of which have been encountered in this book) and yet he had no precise concept of the notion of limit as we understand it. For example the delightful formula $e = \lim_{n \to \infty} \left(1 + \frac{1}{n}\right)^n$ which we met in Chapter 7 was discovered by Euler, but he wrote it as '$e = \left(1 + \frac{1}{i}\right)^i$' where i stood for 'infinity' (and not the square root of minus one in this instance).

The breakthrough to understanding the real nature of the limit came in the nineteenth century and we should honour the contribution of a great trio of mathematicians. The Bohemian mathematician Bernhard Bolzano (1781–1848) realised that a function f is continuous at x if $f(x + h) - f(x)$ can be made as small as you like if h is sufficiently small. Similar ideas were also independently developed by the French mathematician Augustus Louis Cauchy (1789–1857). Cauchy is a very important figure in the history of analysis. We have already met his root and condensation tests for convergence of series in Exercises 6.13

---

[1] In J. le R. d'Alembert and D.Diderot (eds.). Encylopèdie ou dictionnaire raisonnè des sciences, des artes et des metiers. Paris-Neuchâtel-Amsterdam (1765).

and 6.14 (respectively) and the notion of a Cauchy sequence in Chapter 11. He also pioneered the important field of *complex analysis* whereby analytic techniques are extended to sequences and series comprising complex numbers and to complex-valued functions. Indeed this was an essential ingredient for Riemann's work on the analytic continuation of the zeta function that we alluded to in Section 8.3. It was the German mathematician Karl Weierstrass (1815–97) who introduced the modern $\epsilon - \delta$ definition of a limit and was finally able to give rigorous meaning to the notion of sufficiently small.[2] Much of Weierstrass' work was carried out while he was a high school teacher in 1841–56 and didn't become well known until he lectured at Berlin (where he obtained a professorial chair) in 1859. Earlier in 1854 Bernard Riemann (1826–66) (building on work of Cauchy) had established that the integral was essentially a limit of what are now called 'Riemann sums'.

Meanwhile a new front was opened up in analysis when Joseph Fourier (1768–1830) began to study trigonometric series – now called Fourier series in his honour. This opened up a whole range of new problems to solve and led Georg Cantor(1845–1918) to his research on set theory and the infinite.

I haven't said much about the history of sequences and series in this brief chapter. It seems that the first systematic user of sequences in analysis was Cauchy in his *Cours d'Analyse de l École Polytechnique* which was first published in 1821. The first treatise on infinite series appeared a century earlier as an appendix to Jacob Bernoulli's *Ars Conjectandi* which was published in 1713 – eight years after the author's death. The main part of the book is about probability theory but as Bernoulli was one of the first to appreciate, this leads naturally to questions about limits and series (and that's another big story).

So infinite series were being manipulated and their 'sums' calculated long before the rigorous notion of a limit was established. Indeed in Chapter 4 we met the proof that the harmonic series was divergent due to Nicolas Oresme (1323?–82). By the time of the birth of calculus in the seventeenth century, infinite series were being used routinely by mathematicians and important advances were made by John Gregory (1638–75) and Nicolas Mercator (1620–87).

With Weierstrass' definition of the limit established, all the results obtained by earlier mathematicians on calculus and infinite series, products and continued fractions could now be made logically watertight. These subjects had now reached the form in which they are still typically taught in undergraduate mathematics courses. But the work of analysis was far from finished, indeed it could be argued that it had only just begun. As the twentieth century dawned, mathematicians such as Emile Borel (1876–1956) and Henri Lebesgue (1875–1941) explored an abstract concept called *measure*, which enabled a unified treatment of length,

---

[2] Note that Cauchy used ϵ and δ in some of his proofs, although not within his definitions. In fact he was responsible for introducing ϵ into analysis as standing for 'error' (erreur in French) – see Judith Grabiner, 'Who gave you the epsilon? Cauchy and the origins of rigorous calculus'. *American Mathematical Monthly* 90, No. 3, 185–94 (1983).

area, volume and higher-dimensional analogues as well as probability. Out of this work, Lebesgue developed a new theory of integration that was more flexible and powerful than Riemann's. This became a key tool for one of the great developments of twentieth century mathematics – *functional analysis* – which saw analysis and linear algebra unified in a new powerful theory which allowed analytic techniques to be applied within infinite-dimensional spaces. This turned out to be vital for the new physics of quantum theory. Functional analysis is just one of many distinct areas within modern analysis that continues to be developed in the twenty-first century. Analysis is also finding new important application areas within mathematics itself such as fractals, dynamical systems and stochastic processes. Indeed analysis is well established as one of the fundamental strands of mathematics that reaches far into both pure and applied mathematics and there is much to be done in the future.

Further Reading

In this section, I will try to address the question, "What should I read next?" What follows is very much a personal selection and a mixture of classics with the perhaps less well known. I hope no-one is offended by the even longer list of high quality books that I've omitted here. Textbooks are all marked (T).

1. *General, with emphasis on numbers*

(T) R.B.J.T. Allenby, *Numbers and Proofs*, Butterworth-Heinemann (1997).

For those who need a solid grounding in how proof works in mathematics this gives a very thorough and readable textbook account. There is also an introduction to set theory and proof by induction and nice material on prime numbers, integers, rationals, reals and complex numbers.

T. Dantzig, *Number, The Language of Science*, George Allen and Unwin Ltd (1st edition 1936, 4th edition 1962).

This classic account of the history of the number concept from the time of ancient Sumerians and Egyptians to Cantor's theory of the infinite should be accessible to any reader of this book.

J.H. Conway, T.K. Guy, *The Book of Numbers*, Springer Science+Business Media (1996).

If you are fascinated by the patterns that are created by interesting numbers then this is the book for you. You can have fun playing with numbers and shapes, e.g. by thinking about rhombic dodecahedral numbers and also learn about some quite sophisticated topics in number theory such as Ramanujan numbers. The whole book is pervaded by a glorious sense of fun.

M. Gazalé, *Number: From Ahmed to Cantor*, Princeton University Press (2000).

This is another history of the development of numbers and has some similarities with Dantzig's book, but this one contains more mathematics, some of which is quite detailed and demanding. For example there is material on the arithmetic of numbers written in different bases, on continued fractions and an approach to constructing irrational numbers called 'cleavages'.

2. *General, broader scope*

E. Kasner, J. Newman, *Mathematics and the Imagination*, G. Bell and Sons Ltd (1949).

A classic introduction to mathematics for the general reader, but this is also well worth reading for students (and professional mathematicians). It contains a nice account of Cantor's theory of the infinite.

(T) R. Courant, H. Robbins (revised by Ian Stewart), *What is Mathematics?*, Oxford University Press (first edition 1941, second revised edition 1996).

A wonderful introduction to the subject which might be perfect reading on those long hot summer days between school and university. The book covers a lot of ground: numbers, algebra, projective geometry, elementary topology and calculus and includes an introduction to limits and continuity.

L. Gårding, *Encounter with Mathematics*, Springer-Verlag New York Inc (1977).

A survey of mathematics (some quite advanced) for budding mathematicians. This book covers number theory, algebra, geometry, linear algebra, analysis, topology, calculus, probability and applications. If you read this book before first year of university just after Courant and Robbins then you might even be able to teach some of your lecturers a thing or two.

M. Aigner, G.M. Ziegler, *Proofs From the Book* (second edition), Springer-Verlag Berlin, Heidelberg, New York (2001).

Paul Erdös (1913–96) was one of the leading problem-solving mathematicians of the twentieth century and was certainly a unique and remarkable character. To learn more about his life and work you might read his biography, *The Man Who Loved Only Numbers* by Paul Hoffman, Fourth Estate, London (1998). Erdös loved beautiful proofs and whenever he found one he said it should go into 'The Book' (which is perhaps kept by God, who he also referred to as the 'Supreme Fascist'). This lengthy preamble is simply designed to explain the title of this collection of beautiful mathematical proofs in the five selected areas of number theory, geometry, analysis, combinatorics and graph theory. Although they are in some sense 'standard', the proofs that I gave for the irrationality of e and π are strongly influenced by those that appear here.

3. *Textbooks on Analysis – first course level*
These books are listed in increasing order of difficulty, at least as far as I can judge.

(T) R.P. Burn, *Numbers and Functions, Steps into Analysis*, Cambridge University Press (1992).

This covers a standard course in analysis dealing with the 'usual material' of sequences, series, functions, continuity, differentiation and integration. A unique feature is that most of the book consists of guided exercises for the student so you learn by doing rather than just reading.

(T) K.G. Binmore, *Mathematical Analysis, A Straightforward Approach* (second edition), Cambridge University Press (1982).

This is one of the very best textbooks covering a basic first year undergraduate course in analysis. It is well written, nicely explained and has a good collection of exercises. As well as the 'usual material' there are also very welcome chapters on the gamma function and differentiation in higher dimensions.

(T) D. Bressoud, *A Radical Approach to Real Analysis* (second edition), The Mathematical Association of America (2007).

In his introduction the author argues that the 'usual approach' is 'the right way to view analysis but the wrong way to teach it'. He prefers a route that is highly informed by the history of the subject so that students can appreciate how the great masters themselves, such as Euler and Cauchy, struggled with the problems of defining functions and limits. This is a very interesting book and it might be helpful to read this in parallel with taking a traditional course.

(T) T. Tao, *Real Analysis I* (second edition), Hindustan Book Agency (India) (2009).

There is no Nobel prize for mathematics.[1] Instead we have the Fields medals which are awarded every four years to mathematicians under the age of 40 who have made outstanding research contributions to the subject. Terence Tao (b.1975) was awarded a Fields medal in 2006. He has also written a highly innovative two-volume textbook on real analysis. His viewpoint is that a thorough treatment of foundations is required before we encounter the 'usual material' and so he gives an extensive set theoretic treatment of the natural numbers, integers, rational numbers and real numbers (including the construction of the latter using Cauchy sequences) in the first 120 pages. Volume 2 deals with more advanced topics including Fourier series, metric spaces and Lebesgue integration.

(T) D.B. Scott, S.R. Tims, *Mathematical Analysis, An Introduction*, Cambridge University Press (1966).

I include this book (which is sadly out of print) as I first learned analysis from it and it has a special place in my heart. I read it when I was in the sixth form as my maths teacher advised that anyone intending to continue with mathematics at university should study it. How right he was!

(T) S.G. Krantz, *Real Analysis and Foundations* (second edition), Chapman and Hall/CRC (2005).

This is a very wide-ranging book. It starts off with basic material on logic, numbers, sequences and series. Before we complete the 'usual material' of limits, continuity and calculus there is a chapter on topology. The book also contains material on differential equations, Fourier series, multivariate calculus and it ends by giving 'a glimpse of wavelet theory'.

(T) E. Hairer, G. Wanner, *Analysis By Its History*, Undergraduate Texts in Mathematics, Springer (1996).

This takes a similar viewpoint to Bressoud's book in teaching analysis through its history, but it is aimed at a more sophisticated readership.

(T) W. Rudin, *Principles of Mathematical Analysis* (third edition), McGraw-Hill Inc (1976).

This wonderful book is, in my opinion, the all time-winner of the 'best text on elementary analysis' competition. But it is not for the squeamish – after introducing number systems it goes straight into point set topology! I advise you not to read it directly after the current book. First go to one of the texts I listed earlier for the first course and then read this afterwards as the main meal.

---

[1] There is no evidence whatsoever to support the oft-told story that this is because a mathematician had an affair with the mistress of the prize's founder, Alfred Nobel (1833–96).

4. Textbooks on Number Theory – first course level

(T) J. Stopple, *A Primer of Analytic Number Theory*, Cambridge University Press (2003).

Number theory is a vast subject and analytic number theory, i.e. that part of it which relies on analytic methods, is often seen as one of the most challenging and difficult. This beautiful book makes the subject about as accessible as it can be made. The early chapters should even be readable for first year undergraduates.

(T) H. Davenport, *The Higher Arithmetic* (seventh edition), Cambridge University Press (1999).

A classic book that is aimed at general readers although it certainly requires some mathematical sophistication. If you want to learn more about continued fractions, then I recommend looking here as your next step.

(T) G.H. Hardy, E.M. Wright, *An Introduction to the Theory of Numbers* (fifth edition), Oxford University Press (1979).

G.H. Hardy (1877–1947) was one of the greatest UK mathematicians of the twentieth century and he was an analyst par excellence. This is quite an advanced book and yet there are parts of it (such as the first two chapters on prime numbers and the fourth on irrationals) which you should be able to dip into and learn from.

5. Books About Special Numbers

P. Beckmann, *A History of* π, The Golem Press (1971).

This book gives a short history, from the birth of the human race up to the dawn of the computer age, of those parts of mathematics which are associated with the calculation of π or the understanding of its importance. So for example, Chapter 14 gives a five page biography of Euler followed by five pages of his mathematics based around his contributions to π.

E. Maior, *e: The Story of a Number*, Princeton University Press (1995).

The book begins in 1614 with the discovery of logarithms by John Napier (1550–1617) and traces the history of calculus and analysis to the nineteenth century with a particular emphasis on e, e^x and e^{ix}. Among the gems included here you will find an account of an imaginary meeting between Johann Bernoulli and J.S. Bach.

J. Havill, *Gamma, Exploring Euler's Constant*, Princeton University Press (2003).

This book kept me thoroughly absorbed on what looked like being a very boring transatlantic flight (with a screaming 3-year-old sat next to me). If you think that γ is less interesting than e or π, then this will convince you otherwise.

E. Maior, *To Infinity and Beyond*, Princeton University Press (1991).

This book is subtitled 'a cultural history of infinity'. From a mathematical point of view, not only does it cover the analytic approach to infinity which has been featured in the book that you are reading, but also the way infinity appears in geometry (e.g. the 'point at infinity' in projective geometry). You can also read about the role of infinity in aesthetics (e.g. in the paintings of Morris Escher) and the cosmological significance of infinity within modern theories of the universe. A joy to read!

B. Clegg, *A Short History of Infinity*, Constable and Robinson (2003).

This book takes a broad historical view of how our ideas of the infinite have evolved, starting with Zeno's paradoxes of motion and the philosophical views of Aristotle and St Thomas Aquinas and continuing with the discovery of calculus and the mathematical work of Cantor and Godel. Very readable and informative.

6. *History*

M. Kline, *Mathematics in Western Culture*, Penguin (1953).

This is a marvellous book written by one of the twentieth century's greatest popularisers of mathematics. Here you can cross the art/science divide to learn how fifteenth-century painters created the mathematics of projective geometry so that they could incorporate perspective into their paintings.

M. Kline, *Mathematical Thought From Ancient to Modern Times*, Oxford University Press (1972), reprinted in three volumes (1990).

Possibly the most comprehensive history of the subject written so far it begins with the Babylonians about 3000 BCE and ends with developments in set theory and mathematical logic in the early part of the twentieth century.

C.B. Boyer, U.C. Merzbach, *A History of Mathematics* (third edition), J. Wiley and Sons Ltd, (2011).

This is the best known of the histories. It covers similar ground to Kline's three volumes but takes us a little further into the twentieth century. The first edition was solely due to Boyer.

I. Grattan-Guinness, *The Rainbow of Mathematics*, W.W. Norton and Co (1997).

Another history of mathematics from ancient times to the early twentieth century but here there is a greater emphasis on applications to the physical world.

I. James, *Remarkable Mathematicians*, Cambridge University Press (2002).

Contains biographies of 60 mathematicians (including three women) who are allotted about five to seven pages each, starting with Euler in the eighteenth century and finishing with von Neumann in the twentieth century.

J. Fauvel, J. Gray (eds.), *The History of Mathematics: A Reader*, Macmillan Press Ltd (1987).

The great mathematicians in their own words. For example, here you will find some of Newton's writings on the calculus, Cauchy's on convergence of sequences and Cantor's on defining the real numbers.

W. Dunham, *Euler: The Master of Us All*, The Mathematical Association of America (1999).

This is not a biography of Euler. It is an introduction to his wonderful mathematical achievements and comprises a chapter each describing his contributions to number theory, logarithms, infinite series, analytic number theory, complex variables, algebra, geometry and combinatorics.

W. Dunham, *The Calculus Gallery: Masterpieces from Newton to Lebesgue*, Princeton University Press (2005).

> Like the same author's work on Euler, mentioned above, this is really a book about mathematics rather than about its history. He selects eleven mathematicians (including Newton, Cauchy, Riemann, Weierstrass and Lebesgue) and one mathematical family (the Bernoullis) and gives a ten to fifteen page account of their contributions to calculus and/or analysis.

> The MacTutor History of Mathematics Archive (http://www.gap-system.org/history/) is a superb online resource where you can find short biographies of almost any mathematician you can think of (and many more besides).

7. *Easier Reading*

D. Berlinski, *A Tour of the Calculus*, W. Heinemann Ltd. (1996).

> Do read this book. It is on the one hand quite serious mathematics, but on the other hand it is also an awful lot of fun. Need I say more?

There is no need to be restricted to reading books. There are also a number of journals aimed at teachers and students of mathematics which often contain interesting articles about analysis, calculus and related matters and sometimes these don't require greater knowledge or sophistication than you needed to read this book. Those you might look at include *The American Mathematical Monthly, Mathematical Spectrum, The Mathematical Gazette* and *Mathematics Magazine*.

Appendices

Appendix 1: The Binomial Theorem

We've used the *binomial theorem* on several occasions in this book. In this appendix we'll discuss it and sketch a proof. Let x and y be real numbers. We want an expression for $(x+y)^n$ where n is an arbitrary natural number.

Now we can do some basic algebra to calculate

$$(x+y)^2 = (x+y)(x+y) = x^2 + 2xy + y^2,$$

$$(x+y)^3 = (x+y)(x+y)^2 = x^3 + 3x^2y + 3xy^2 + y^3,$$

$$(x+y)^4 = (x+y)(x+y)^3 = x^4 + 4x^3y + 6x^2y^2 + 4xy^3 + y^4,$$

but what about a general formula? We need to expand

$$(x+y)^n = \underbrace{(x+y)(x+y)\cdots(x+y)}_{n \text{ times}}.$$

From the form of the brackets, it is clear that,[1]

$$(x+y)^n = c_{0,n}x^n + c_{1,n}x^{n-1}y + c_{2,n}x^{n-2}y^2 + \cdots + c_{n-1,n}xy^{n-1} + c_{n,n}y^n,$$

where $c_{0,n}, c_{1,n}, \ldots, c_{n,n}$ are natural numbers. So the work required is to identify these numbers. We'll argue in a *combinatorial* manner. The generic term involving x^ry^{n-r} requires us to extract r of the xs and $(n-r)$ of the ys from the n brackets we are multiplying. This is equivalent to the problem of having n containers each containing a ball and having to choose r balls from these without replacing balls that have been removed. In how many ways can this be done? Well there are n ways of choosing the first, $n-1$ ways of choosing the second, $n-2$ ways of choosing the third and we keep going until we get to the last. There are $n-(r-1)$ ways of choosing this one. So altogether the number of ways of choosing the balls is

$$n(n-1)(n-2)\cdots(n-r+1) = \frac{n!}{(n-r)!}.$$

---

[1] This is more along the lines of a convincing argument than a fully rigorous proof, for that we need the technique of mathematical induction.

But these balls have been chosen in a particular order and that should be irrelevant. The total number of ways of ordering the r balls is $r!$ and so we conclude that

$$c_{r,n} = \frac{n!}{(n-r)!r!}.$$

We usually use the notation $\binom{n}{r}$ instead of $c_{r,n}$ and these numbers are called *binomial coefficients*. Since $\binom{n}{0} = \binom{n}{n} = 1$ we can succinctly write:

Theorem A.1 (The Binomial Theorem). If x and y are arbitrary real numbers and n is a natural number:

$$(x+y)^n = \sum_{r=0}^{n} \binom{n}{r} x^r y^{n-r}.$$

You should check that $\binom{n}{1} = \binom{n}{n-1} = n$ and more generally $\binom{n}{r} = \binom{n}{n-r}$. The latter allows us to treat x and y symmetrically in the binomial theorem so that we also have

$$(x+y)^n = \sum_{r=0}^{n} \binom{n}{r} x^{n-r} y^r.$$

As an exercise you should compute $\binom{6}{2} = 15$ and $\binom{6}{3} = 20$ and deduce that

$$(x+y)^6 = x^6 + 6x^5y + 15x^4y^2 + 20x^3y^3 + 15x^2y^4 + 6xy^5 + y^6,$$

without having to multiply out any brackets.

The binomial theorem was apparently first discovered in 1664 or 1665 by the great Isaac Newton and communicated by him in two letters sent in 1676 to Henry Oldenberg who was secretary of the Royal Society. The pattern played by the binomial coefficients and displayed below (where the right-hand side indicates the binomial expansion that these come from) is called *Pascal's triangle* in honour of the mathematician and philosopher Blaise Pascal (1623–62). He was apparently the first in the Western world to notice that each coefficient appearing in the triangle is the sum of the two numbers immediately above it which are placed to the right and to the left.

$$
\begin{array}{ccccccccccc}
 & & & & & 1 & & & & & & (x+y)^0 \\
 & & & & 1 & & 1 & & & & & (x+y)^1 \\
 & & & 1 & & 2 & & 1 & & & & (x+y)^2 \\
 & & 1 & & 3 & & 3 & & 1 & & & (x+y)^3 \\
 & 1 & & 4 & & 6 & & 4 & & 1 & & (x+y)^4 \\
1 & & 5 & & 10 & & 10 & & 5 & & 1 & (x+y)^5 \\
\end{array}
$$

etc.

You are invited to try to prove the formula that expresses this:

$$\binom{n+1}{k} = \binom{n}{k} + \binom{n}{k-1}.$$

It only requires basic algebra.

If we take $y = 1$ in the binomial theorem and cancel the coefficients down as far as is possible we get

$$(1+x)^n = \sum_{r=0}^{n} \binom{n}{r} x^r$$

$$= 1 + nx + \frac{1}{2}n(n-1)x^2 + \frac{1}{6}n(n-1)(n-2)x^2 + \cdots + x^n.$$

Newton was the first to consider what happens when n is allowed to take more general values. In fact if c is an arbitrary real number and $-1 < x < 1$, then $(1+x)^c$ has meaning as the sum of a convergent infinite series – called the *binomial series*, and we have

$$(1+x)^c = 1 + cx + \frac{1}{2}c(c-1)x^2 + \frac{1}{6}c(c-1)(c-2)x^2 + \cdots$$

In the special case $c = -1$ we get

$$(1+x)^{-1} = 1 - x + x^2 - x^3 + \cdots$$

and this was used in Section 7.2 to obtain Gregory's series from which we deduced a series expansion for π.

Appendix 2: The Language of Set Theory

With the exception of Chapter 11, this book has not used set theory at all. Indeed for most of the book is wasn't necessary. However it is an essential tool for going further in analysis and a brief account of it might help to make other textbooks more accessible. This short appendix gives an introduction to this important area of mathematics.

A set is a mathematical way of representing a collection of 'objects'. It doesn't matter what these objects are. They could be mundane things or mathematical symbols. Let's begin with an example. The colours of the rainbow are red, orange, yellow, green, blue, indigo and violet. We will collect them together as a set which we will denote by C for 'colours'. We write this set

$$C = \{\text{red, orange, yellow, green, blue, indigo,violet}\}.$$

On the left-hand side of the equals sign we have the name we've given our set. On the right-hand side we have the list of the *elements* or *members* of the set. These are separated by commas. The braces { and } signal the beginning and end of the set (respectively). In a set the order in which we write the list is irrelevant.[2] I have chosen to write the rainbow colours in the familiar ordering whereby wavelength decreases from left to right but I could just as easily have written e.g.

$$C = \{\text{yellow, orange, violet, blue, red, indigo, green}\}.$$

It is helpful to have a mathematical way of indicating set membership and the symbol $\in$ fulfils this role. So if we want to write succinctly that 'yellow is a member of then set C' we just say yellow $\in C$. But although black is a perfectly good colour, it does not appear in the rainbow and so is not a member of C. We denote this by black $\notin C$.

Set theory is of universal use in mathematics. In analysis (at least at the level of this book) we usually want to consider sets whose elements are numbers. For example we might want to consider the set of all integers which lie strictly between -5 and 2 and this is the set

$$S_1 = \{-4, -3, -2, -1, 0, 1\}.$$

This set has precisely six elements but many important sets have an infinite number of elements. For example, the set of all natural numbers is universally denoted by the symbol $\mathbb{N}$. We cannot write this as a full list but we can at least indicate how the list starts:

$$\mathbb{N} = \{1, 2, 3, 4, 5, \ldots\}.$$

The sets of all integers, rational numbers and real numbers are written $\mathbb{Z}$, $\mathbb{Q}$ and $\mathbb{R}$ (respectively). We can write

$$\mathbb{Z} = \{\ldots, -3, -2, -1, 0, 1, 2, 3, 4, \ldots\}.$$

Sometimes we can write a set in a succinct way by using a formula or relation that generates all the elements in the list. For example consider the set S_1 that we defined above. We can also write this

$$S_1 = \{x \in \mathbb{Z}; -4 \leq x \leq 1\}$$

or equivalently

$$S_1 = \{x \in \mathbb{Z}; -5 < x < 2\}.$$

The semicolon here stands for 'such that' or 'for which', so the meaning of the right-hand side in the first of these expressions for S_1 is precisely that we have the set of all integers x for which x lies between -4 and 1.

---

[2] There is a notion of an *ordered set* which is important in the axiomatic approach to mathematical analysis, but we won't pursue that direction here.

The rational numbers can be given a nice description from this point of view:

$$\mathbb{Q} = \left\{ x \in \mathbb{R}; x = \frac{p}{q}, p \in \mathbb{Z}, q \in \mathbb{N} \right\}.$$

Relationships between sets are important. We say that a set A is a subset of a set B if every element of A is also an element of B. In this case we write $A \subseteq B$. The line underneath $\subset$ indicates that it may well be that A and B are the same set. If we know that $A \subseteq B$ but that there are elements of B that are not in A then we write $A \subset B$ and say that A is a proper subset of B. The relationship between $\subseteq$ and $\subset$ is analogous to that between $\leq$ and $<$ in the study of inequalities. For example, considering the sets we've discussed above, we have $S_1 \subset \mathbb{Z}$ and

$$\mathbb{N} \subset \mathbb{Z} \subset \mathbb{Q} \subset \mathbb{R}.$$

In Chapter 8 we introduced the complex numbers and we can also write the set of all of these as

$$\mathbb{C} = \{x + iy; x, y \in \mathbb{R}\},$$

where $i = \sqrt{-1}$. As every real number x can be written in the form $x + i0$ we see that $\mathbb{R} \subset \mathbb{C}$.

Intervals are examples of subsets of $\mathbb{R}$, e.g. $[a, b] = \{x \in \mathbb{R}; a \leq x \leq b\}$.

In many areas of mathematics where set theory is used we may often identify a *universal set* which has the property that all other sets that we consider are subsets of it. For much of this book we have been concerned with the subject of real analysis and here the universal set is $\mathbb{R}$. We have briefly touched on the subject of complex analysis where the universal set is $\mathbb{C}$. For much elementary work on the theory of numbers we might take $\mathbb{N}$ to be our universal set.

Now suppose that X is the universal set and that A and B are subsets of X. There are useful ways in which we can form new sets from A and B. The most basic of these are called the complement, union and intersection. The complement is usually denoted A^c or $\overline{A}$. It is defined to be the set of all elements of X that are not in A.

$$A^c = \{x \in X; x \neq A\}.$$

The union of A and B is written $A \cup B$. It is the set of all elements of X which are either members of A, members of B or members of both A and B. The intersection of A and B is denoted $A \cap B$. It is the set of all elements of X which are members of both A and B.

A simple example may help to clarify these definitions:

Let $X = \{1, 2, 3, 4, 5, 6, 7, 8, 9, 10\}$ and define $A = \{2, 4, 6, 8, 10\}, B = \{3, 6, 8, 9\}$ and $C = \{1, 6, 10\}$.

Then $A^c = \{1, 3, 5, 7, 9\}, B \cup C = \{1, 3, 6, 8, 9, 10\}$ and $B \cap C = \{6\}$.

The *empty set* Ø is defined to be X^c. It is a set that contains no elements and it can be very useful for expressing that which can never be, e.g.

$$\emptyset = \{x \in \mathbb{N}; \ x \text{ is prime}, \ x \text{ is even and } x > 2\}.$$

As indicated above, the union is defined using the 'inclusive or' (i.e. in A or in B or in both) rather than the 'exclusive or' of ordinary language. Whenever I write 'or' from now on in this section I will mean this inclusive sense. Now suppose that $A_1, A_2, \ldots, A_n$ are a finite number of subsets of X. Then we can define the intersection $A_1 \cap A_2 \cap \cdots \cap A_n$ to be the subset of all elements of X that are in every set A_j where $1 \leq j \leq n$ and the union $A_1 \cup A_2 \cup \cdots \cup A_n$ to be the subset comprising those elements of X that are in at least one of the A_js. You should think about how the phrase 'at least one' is exactly equivalent to being a member of A_1 or of A_2 or of etc. Sometimes we use a similar notation to the sigma notation for sums to express these unions and intersections, so we have $\bigcap_{i=1}^{n} A_i = A_1 \cap A_2 \cap \cdots \cap A_n$ and $\bigcup_{i=1}^{n} A_i = A_1 \cup A_2 \cup \cdots \cup A_n$. These ideas extend to countably or even uncountably many sets and the union over the former appears in Theorem 11.3.1. To see how this works, let's suppose we have a sequence of sets (A_n). We define $\bigcap_{n=1}^{\infty} A_n$ to be the subset of X comprising elements that belong to every member of the sequence and $\bigcup_{n=1}^{\infty} A_n$ to be the subset of X comprising elements that belong to at least one member of the sequence. For example if we take $X = \mathbb{R}$ then you may want to think about why $[0, \infty) = \bigcup_{n=1}^{\infty} [n-1, n]$ and $[0, 1] = \bigcup_{n=2}^{\infty} \left(\frac{1}{n}, 1 \right]$ There is a great deal more that we could say about set theory, but this is a good place to stop as we have more than covered all that is needed for the rather minimal use that we've made of it in the main part of the text.

Appendix 3: Proof by Mathematical Induction

In the introduction I pointed out that I would avoid using proof by mathematical induction in this book. This appendix is included for those of you you want to know what this is, how it works and what it can do for us. Be aware that this technique is indispensable for anyone doing an undergraduate degree that involves sophisticated mathematics.

The context for mathematical induction is a sequence of propositions (P_n) which start at some nonnegative integer n_0. So the sequence begins $P_{n_0}, P_{n_0+1}, P_{n_0+2}, P_{n_0+3}, \ldots$. Quite often we find that n_0 is 1 or 0 but this is not essential. It is however crucial for this particular method of proof that the total number of propositions is (countably) infinite. Mathematical induction is a device that gives a method for simultaneously proving the validity of all the propositions P_n provided that two steps are carried out successfully:

Step 1 (The Initial Step.) Prove that P_{n_0} is true.

186

Step 2 (The Inductive Step.) Prove that if for some arbitrary n, P_n is true then P_{n+1} is true.

If steps 1 and 2 can both be carried out successfully then the principle of mathematical induction states that P_n is true for every $n \geq n_0$. We will take for granted that this principle holds. You should be able to see the logic that underlies it. Step one tells us that P_{n_0} is true. Now apply step 2 with $n = n_0$ to deduce that P_{n_0+1} is true. Then apply step 2 again with $n = n_0 + 1$ to see that P_{n_0+2} is true, and so on, ad infinitum. The proposition P_n (where n is arbitrary) is sometimes called the *inductive hypothesis* in the context of mathematical induction.

Example A.1: In Chapter 6, we met the famous formula for the sum of the first n natural numbers. Two proofs were given of this result, one allegedly due to Gauss as a schoolboy and the other using the formula for the sum of an arithmetic progression. We'll now give a proof by mathematical induction. Here $n_0 = 1$ and P_n is the proposition

$$1 + 2 + \cdots + n = \frac{1}{2}n(n+1).$$

Step 1. When $n = 1$ the left-hand side of the formula is 1 and the right-hand side is $\frac{1}{2}1(1+1) = 1$ so clearly P_1 holds.

Step 2. Assume the result holds for some n then

$$1 + 2 + \cdots + n + n + 1 = (1 + 2 + \cdots + n) + n + 1$$
$$= \frac{1}{2}n(n+1) + (n+1)$$
$$= (n+1)\left(\frac{n}{2}+1\right)$$
$$= \frac{1}{2}(n+1)(n+2),$$

and so we see that P_{n+1} is true.

Step 1 and Step 2 have both been verified and so we can assert that the required result holds for all natural numbers n by mathematical induction. As an exercise you may like to try to use mathematical induction to prove that

$$1 + 2^2 + 3^2 + \cdots + n^2 = \frac{1}{6}n(n+1)(2n+1),$$

$$\text{and } 1 + 2^3 + 3^3 + \cdots + n^3 = \frac{1}{4}n^2(n+1)^2.$$

In analysis, proof by induction can be a useful tool for establishing inequalities.

Example A.2: In Section 5.3 we met the sequence (a_n) defined by

$$a_1 = 1 \text{ and } a_{n+1} = \sqrt{1 + a_n} \text{ for } n = 2, 3, 4, \ldots.$$

We used a proof by contradiction there to establish the bound $a_n \le 2$ for all n. Now let's proceed by induction. Here we again have $n_0 = 1$ and Step 1 is obvious. For Step 2, assume the bound holds for some n. Then $a_{n+1} \le \sqrt{1 + 2} = \sqrt{3} < 2$. So the result holds for $n + 1$ and so is true for all n by induction.

Example A.3: Here we return to Problem 6 at the end of Chapter 3. There you were invited to prove the inequality $(a + b)^2 \le 2(a^2 + b^2)$ and you were also asked to guess what shape the inequality takes when the two real numbers a and b are replaced by n real numbers. Here's the answer

$$(a_1 + a_2 + \cdots + a_n)^2 \le n(a_1^2 + a_2^2 + \cdots + a_n^2).$$

We will prove this by using mathematical induction. Here we take $n_0 = 2$ and observe that Step 1 was exactly what you established in Problem 6 in Chapter 3. We have to work a little to carry out the inductive step. Assume that the required inequality holds for some n. Then

$$(a_1 + a_2 + \cdots + a_n + a_{n+1})^2$$
$$= [(a_1 + a_2 + \cdots + a_n) + a_{n+1}]^2$$
$$= (a_1 + a_2 + \cdots + a_n)^2 + 2a_{n+1}(a_1 + a_2 + \cdots + a_n) + a_{n+1}^2$$
$$\le n(a_1^2 + a_2^2 + \cdots + a_n^2) + 2a_{n+1}a_1 + 2a_{n+1}a_2 + \cdots + 2a_{n+1}a_n + a_{n+1}^2.$$

Now use the fact that $2a_{n+1}a_j \le a_{n+1}^2 + a_j^2$ for each $1 \le j \le n$,[3] to find

$$(a_1 + a_2 + \cdots + a_n + a_{n+1})^2$$
$$\le n(a_1^2 + a_2^2 + \cdots + a_n^2) + na_{n+1}^2 + a_1^2 + a_2^2 + \cdots + a_n^2 + a_{n+1}^2$$
$$= (n + 1)(a_1^2 + a_2^2 + \cdots + a_n^2 + a_{n+1}^2),$$

and the required result follows by mathematical induction.

Appendix 4: The Algebra of Numbers

Suppose that a, b and c are arbitrary natural numbers (respectively, integers, rational numbers, real numbers, complex numbers) then the sum $a + b$ and

---

[3] This follows from the fact that $(a_{n+1} - a_j)^2 \ge 0$ for $1 \le j \le n$ and was already used in Problem 6 of Chapter 3.

product ab are also natural numbers (respectively, integers, rational numbers, real numbers, complex numbers) and the following always hold

- *Commutative Law of Addition*

$$a + b = b + a,$$

- *Associative Law of Addition*

$$(a + b) + c = a + (b + c),$$

- *Commutative Law of Multiplication*

$$ab = ba,$$

- *Associative Law of Multiplication*

$$a(bc) = (ab)c,$$

- *Distributive Law of Multiplication Over Addition*

$$a(b + c) = ab + ac.$$

The integers (respectively, rational numbers, real numbers, complex numbers) have the property that each element a has an *additive inverse* $-a$, i.e.

$$a + (-a) = (-a) + a = 0,$$

and the set of all integers (respectively, rational numbers, real numbers, complex numbers) is an example of an important algebraic structure called a *ring*. The rational numbers (respectively real numbers, complex numbers) have the property that each non-zero element a has a *multiplicative inverse* $\frac{1}{a}$, i.e.

$$a.\frac{1}{a} = \frac{1}{a}.a = 1,$$

and the set of all rational numbers (respectively real numbers, complex numbers) is an example of another important algebraic structure called a *field*. The sets of rational numbers and real numbers (but not the set of complex numbers) are both provided with an *order relation* $\leq$ and these are called *ordered fields*. Both the sets of real numbers and the complex numbers (but not the set of rational numbers) have the completeness property (see Chapter 11) that every Cauchy sequence converges to a limit therein. The set of real numbers is the unique *complete ordered field*. If you are curious to know what a ring or a field is then you can search for these concepts on Wikipedia, but it's even better to read an introductory text on abstract algebra.

Hints and Solutions to Selected Exercises

Exercise 1.3 Write $m = 2k - 1$ and $n = 2l$.

Exercise 1.4 Add together m, $m + 1$, $m + 2$ and $m + 3$ and show that the answer is of the form $2p$.

Exercise 1.5 (a) If n is odd then we can write $n = 2p - 1$ but either p is even ($p = 2m$) or p is odd ($p = 2m + 1$). Every odd number greater than 3 of the form $4m - 1$ can be written in the form $4k + 3$ by writing $m = k + 1$.

Exercise 1.7 (a)

$$
\begin{array}{llll}
p = 2, & 2^2 - 1 = 3 & \text{which is prime} \\
p = 3, & 2^3 - 1 = 7 & \text{which is prime} \\
p = 4, & 2^4 - 1 = 15 = 3 \times 5 & \text{which is composite} \\
\text{But for} \quad p = 5, & 2^5 - 1 = 31 & \text{which is prime}
\end{array}
$$

In fact $p = 7, 13, 17$ and 19 also yield Mersenne primes, but as of October 2009 only 47 of these have been found, the largest of which corresponds to $p = 43112609$ (see http://en.wikipedia.org/wiki(Mersenne_prime).

(b)

$$
\begin{array}{lll}
n = 0, & 2^{2^0} + 1 = 3 & \text{which is prime} \\
n = 1, & 2^{2^1} + 1 = 5 & \text{which is prime} \\
n = 2, & 2^{2^2} + 1 = 17 = 3 & \text{which is prime} \\
n = 3, & 2^{2^3} + 1 = 257 & \text{which is prime} \\
n = 4, & 2^{2^4} + 1 = 65537 & \text{which is prime.}
\end{array}
$$

No other Fermat numbers are known. When $n = 5$, $2^{2^5} + 1 = 4294967297 = 641 \times 6700417$.

Exercise 1.8 Every number between 1 and 40 (inclusive) generates a prime number but $n = 41$ yields a number that is composite.

Chapter 2

Exercise 2.1 0.1101

Exercise 2.2 $0.\dot{0}7692\dot{3}$

Exercise 2.3 (b) $\frac{1264}{2475}$

Exercise 2.4 First show that $5\frac{10}{61} = \frac{1890}{366}$ and $5\frac{1}{6} = \frac{1891}{366}$.

Exercise 2.5 (a) $1296 = 36^2$.

Exercise 2.6 e.g. $x = \sqrt{2}$ and $y = \frac{1}{\sqrt{2}}$.

Exercise 2.9 The answer is yes and here is a 'non-constructive' proof. $\sqrt{2}^{\sqrt{2}}$ is either rational or irrational. If it is rational choose $a = b = \sqrt{2}$ and if it is irrational choose $a = \sqrt{2}^{\sqrt{2}}$ and $b = \sqrt{2}$. [In fact in can be shown that $\sqrt{2}^{\sqrt{2}}$ is irrational but this needs some very advanced techniques.]

Exercise 2.10 It is irrational as it is not periodic or even eventually so. For an example of an appropriate rational number take
$0.123456789101\dot{2}\dot{0}$
and for an appropriate irrational number:
$0.123456789101124262830323\underline{4}\cdots$.

Chapter 3

Exercise 3.2 $bd - ac = bd - bc + bc - ac = b(d-c) + c(b-a) \geq 0$.

Exercise 3.3 $x < -3$ or $-1 \leq x \leq 2$.

Exercise 3.4 $0 < x < 2$.

Exercise 3.7 $(1+x)^r = 1 + rx + \frac{1}{2}r(r-1)x^2 + \frac{1}{6}r(r-1)(r-2)x^3 + \cdots + x^r \geq 1 + rx$.

Chapter 4

Exercise 4.1 Limit is 1. (a) (i) 30, (ii) 300, (iii) 3000, (iv) 30000, (v) 30000000000.
(b) Take n_0 to be the smallest natural number so that $n_0 + 1$ exceeds $\frac{3}{\epsilon}$.

Exercise 4.2 Limits are (a) 1, (b) 0, (c) 0, (d) 0.

Exercise 4.4 Use $||x_n| - |x|| \leq |x_n - x|$. Converse is false – e.g consider $(-1)^n$.

Exercise 4.6 Limits are (a) 6, (b) 1, (c) $\frac{2}{5}$, (d) $\frac{1}{2}$, (e) 0.
Hint for (e) – multiply top and bottom by $\sqrt{n+1} + \sqrt{n}$.

Exercise 4.8 (b) We must show that $\frac{1}{n} < \epsilon$ for all $n > N$. But this means that $n > \frac{1}{\epsilon}$ for such n. Suppose that such an N can be found and choose $\epsilon = \frac{1}{N+2}$. Now take $n = N + 1$. What do you find?

Exercise 4.11 $c_n - a_n \geq b_n - a_n \geq 0$.

Exercise 4.13 If $x_n \to l$ then $x_{n+1} \to l$ as $n \to \infty$. (i) $\frac{1}{n!}$, (ii) $\frac{1}{2^n}$.

Exercise 4.15 Let (a_{n_r}) be a subsequence. Given $\epsilon > 0$ there exists N such that if $n > N$ then $|a_n - l| < \epsilon$. Choose a natural number R to be such that $n_r > N$ whenever $r > R$ and the result follows.

Chapter 5

Exercise 5.2 (a) sup 3, inf -2, increasing to 3, (b) sup 2, inf 1, decreasing to 1, (c) inf 2, increasing, diverges to $+\infty$, (d) inf -1, sup $\frac{1}{2}$, oscillates finitely, (e) inf 0, sup 2, converges to 0.

Exercise 5.3 (a) Write $A = \sup(a_n)$ and $B = \sup(b_n)$. As $a_n \leq A$ and $b_n \leq B$ for all n it follows from Exercise 3.2 that $a_n b_n \leq AB$ for all n. So AB is an upper bound for the sequence $(a_n b_n)$ and so $AB \geq \sup(a_n b_n)$. A counter-example is $a_n = 1 + \frac{1}{n}$, $b_n = 1 - \frac{1}{n}$.

Exercise 5.4 $\inf(a_n)\inf(b_n) \leq \inf(a_n b_n)$. A counter-example for the first inequality in the question is as in Exercise 5.3 (a) above, for the second take $a_n = n$, $b_n = \frac{1}{n}$.

Exercise 5.6 The sequence is monotonic decreasing to 2.

Exercise 5.7 (a) The Theorem of the Means is a useful tool. (c) Deduce that (a_n) is monotonic decreasing and bounded below and that (b_n) is monotonic increasing and bounded above. To show that they have the same limit, first establish that $a_{n+1} - b_{n+1} \leq \frac{1}{2^n}(a - b)$.

Exercise 5.10 (a) By the result of (8) $a_{n_k} \leq l$ for all k. Now for any $n_k \leq n \leq n_{k+1}$ we have $a_n \leq a_{n_{k+1}} \leq l$ and the result follows. (b) Given any $\epsilon > 0$ we can find K such that if $k > K$ then $l - \epsilon < a_{n_k} < l + \epsilon$. Then for $n > n_k$, $l - \epsilon < a_{n_k} < a_n \leq l < l + \epsilon$ and the result follows.

Exercise 5.12 (i) lim sup 1, lim inf -1, (ii) lim sup 0, lim inf 0, (iii) lim sup 1, lim inf -1.

Chapter 6

Exercise 6.1 (a) Converges to 150, (b) diverges, (c) converges to 2, (d) converges to $\frac{3}{4}$, (e) converges to $\frac{1}{1-\sin(\theta)}$ for all values of θ except for $\frac{\pi}{2}$ and $\frac{3\pi}{2}$ where the series diverges.

Exercise 6.3 Using (6.7.14) $\sum\limits_{n=1}^{\infty} a_n = \sum\limits_{n=1}^{N-1} a_n + \sum\limits_{n=N}^{\infty} a_n$ and then by algebra of limits,

$$\lim_{N\to\infty} \sum_{n=N}^{\infty} a_n = \lim_{N\to\infty} \left(\sum_{n=1}^{\infty} a_n - \sum_{n=1}^{N-1} a_n \right) = \sum_{n=1}^{\infty} a_n - \sum_{n=1}^{\infty} a_n = 0.$$

Exercise 6.4 (a) $\frac{17}{99}$.

Exercise 6.5 (a) Does $\lim_{n\to\infty} a_n = 0$ in either (i) or (ii)?

Exercise 6.6 (a), (b), (d) converge – (c), (e) diverge.

Exercise 6.7 (a), (b), (c) all converge (for (c) note that this is because $\frac{1}{e} < 1$), (d) diverges.

Exercise 6.8 (a) converges (ratio test if you know about e or root test, for which see Exercise 6.13) (b), (c) diverge (both comparison test), (d) converges (ratio test).

Exercise 6.10 Use the Theorem of the Means to show that
$$\sqrt{a_n a_{n+1}} \le \tfrac{1}{2}(a_n + a_{n+1}).$$

Exercise 6.15 (a) All three series are convergent by the Leibniz test. (b)(i) is not absolutely convergent, but (ii) and (iii) both are – you should use the comparison test in all three cases. For (ii), note first that for $n > 2$, $3n^2 - 2n > 3n^2 - 6n = 3n(n-2) > 3(n-2)^2$ so $\frac{1}{3n^2-2n} < \frac{1}{3(n-2)^2}$.

Exercise 6.16 (a) False – e.g $a_n = \frac{1}{n}$

(b) True. If $\sum\limits_{n=0}^{\infty} |a_n|$ converges then $\lim_{n\to\infty} a_n = 0$, hence we can find n_0 such that $n > n_0 \Rightarrow |a_n| < 1$. But then $a_n^2 < |a_n|$ for all $n > n_0$ and the result follows from the comparison test.

Exercise 6.17 e.g. $a_n = (-1)^n \frac{1}{\sqrt{n}}$.

Exercise 6.19 (a) Since the series converges to s say, then given $\epsilon > 0$ there exists N such that if $n > N$ then $|s_n - s| < \frac{\epsilon}{2}$. Then

$$|s'_n - s| = \left| \frac{s_1 + s_2 + \cdots + s_n}{n} - s \right|$$

$$\leq \left| \frac{s_1 + s_2 + \cdots + s_N - Ns}{n} \right| + \left| \frac{s_{N+1} + \cdots + s_n - (n-N)s}{n} \right|$$

$$\leq \left| \frac{s_1 + s_2 + \cdots + s_N - Ns}{n} \right| + \frac{|s_{N+1} - s|}{n} + \frac{|s_{N+2} - s|}{n} + \cdots$$

$$+ \frac{|s_n - s|}{n}$$

$$\leq \left| \frac{s_1 + s_2 + \cdots + s_N - Ns}{n} \right| + \frac{(n-N)\epsilon}{2n}$$

$$< \left| \frac{s_1 + s_2 + \cdots + s_N - Ns}{n} \right| + \frac{\epsilon}{2},$$

and the result follows since $\lim_{n \to \infty} \frac{1}{n} = 0$ and so for sufficiently large n, $\left| \frac{s_1 + s_2 + \cdots + s_N - Ns}{n} \right| < \frac{\epsilon}{2}$ (but you should fill in the details to make this precise). (b) $s_n = \frac{1}{2}(1 - (-1)^n)$. If n is even then $s'_n = \frac{1}{n} \left(\frac{n}{2} \right) = \frac{1}{2}$ and if n is odd, $s'_n = \frac{1}{n} \left(\frac{n+1}{2} \right) \to \frac{1}{2}$ as $n \to \infty$, from which the result follows.

Index

$$\frac{120}{200}$$